THE
WASTE
MANAGERS

THE
WASTE
MANAGERS

The team that built **WASTE MANAGEMENT, INC.**,
one of North America's most important companies

WILLIAM J. PLUNKETT

REPUBLIC
BOOK PUBLISHERS

THE WASTE MANAGERS

FIRST EDITION

Copyright 2022 William J. Plunkett

ISBN: 978-1-64572-070-6 (Hardcover)
ISBN: 978-1-64572-071-3 (Ebook)

For inquiries about volume orders, please contact:
Republic Book Publishers
27 West 20th Street
Suite 1103
New York, NY 10011
editor@republicbookpublishers.com

Published in the United States by Republic Book Publishers
Distributed by Independent Publishers Group
www.ipgbook.com

Book designed by Mark Karis
Printed in the United States of America

Contents

CONTENTS

Former Waste Management CEO Dean Buntrock reviews the prospectus detailing the company's June 1971 initial public offering of stock. *Photo credit: Buntrock family archive*

FOREWORD

BY DEAN BUNTROCK

I knew I wanted to do something special to celebrate Waste Management's fiftieth birthday, something that would be meaningful to the thousands of men and women who made their careers at the company—and something that would last long after the weekend reunion I was planning had simply become a memory. And so I asked Bill Plunkett, a colleague who had come to Waste Management from the *Chicago Tribune* and Tribune Company, to write this book about the company and the men and women who made the remarkable Waste Management story possible.

I began thinking about commissioning this book three years ago when Waste Management, Inc., which now generates nearly $18 billion in annual revenue, had grown 3,000 times larger than when I formed the company in 1968 and well over 20,000 times larger than when I began managing the family trash hauling business in 1956 following the untimely passing of my father-in-law. I was newly married then and barely a year out of college. Now, sixty-six years later, I sometimes shake my head in wonderment when I contemplate what we built at Waste Management and the thousands of careers that blossomed during the forty years that I had the good fortune to lead this great American company. I did not start out seeking to build a great American company. Like most young men, I wanted to build a secure life for myself and a loving family. I never could have anticipated the unforeseen circumstances that were to emerge or the crucial decisions I would be called upon to make.

Each decision we make, the good ones and the poor ones, can take us in one direction or another and rarely in a perfectly straight line. The Waste Management story is, to a great extent, the story of my life and the lives of the hardworking and talented men and women who joined me on an incredible journey. Let me tell you about it.

THE ROAD TO WASTE MANAGEMENT

I was born in Columbia, South Dakota, during a dark time in American history. The US was in the depths of the most profound economic depression ever recorded. It was a stressful time for most Americans, who had little reason to feel secure or confident about their futures. Their livelihoods hung in the balance day by day. The greatest drought in American history was about to descend on a vast swath of North America, and, as one writer observed, "it was a hard time to be alive in South Dakota."

Columbia, not far from Aberdeen, was home to about two hundred and fifty people. Fewer than one hundred and forty souls call Columbia home today. Columbia lacked the hustle and bustle of typical large American cities and suburban centers that many major businesses and hard-driving businessmen called home. Still, the town, and the good Lord, provided me with experiences and values that have served me incredibly well throughout my life.

My family had a modest farm implement store, and from the time I was a young teenager, I worked by my father's side. When I wasn't in school, I was at the store helping and mostly learning. It was my favorite time. I came to understand at a pretty early age that business was simply about providing a product or a service that others valued. I learned more about business as a teenager in Columbia than I did later as a student away in college.

The hours were long, and the work was demanding. The workday began early and ended when the work was done. That was my routine, except when I was in school. We worked all day every Saturday. Sunday was for church, counting our blessings and giving thanks for those blessings.

Upon graduating from high school, I went to St. Olaf College, a relatively small Lutheran school in Northfield, Minnesota, that my mother had selected for me. She let me know before I entered grade school that someday I was going to go to college. At St. Olaf, I was exposed to a wide range of new subjects and new things to think about. I met other young men and women from different parts of the country, including some veterans who were there thanks to the GI Bill. I felt like a fish out of water there. The small high school I had graduated from with seven other seniors covered the basics well enough but really didn't prepare me for the faster pace of college learning nor how to study for the wide range of subject matters that I was introduced to. Fortunately, the professors at St. Olaf were patient and dedicated and gave me, and many other students, the time we needed to keep up with the demands of college.

My college education, however, was interrupted by circumstances—the kind that often interrupt and sometimes alter the course of one's life. During my junior year, a debilitating illness struck my father, and I had to return to Columbia to keep the business going until I could arrange for it to be sold. Meanwhile, the Korean War was raging, and, while home, I received my draft notice. I had up to one year to report for duty or to report whenever the business was sold, whichever came first. Ten months later, I was on a train heading to Fort Leonard Wood in the Missouri Ozarks.

The Army was a thoroughly positive and important growing experience for me. After basic training, I was trained in field engineering, gained worthwhile experience in training new enlistees, and later, was sent to the Army Finance School at Fort Benjamin Harrison in Indianapolis, Indiana.

When the war ended twenty-one months later, I was discharged in order to attend summer school before returning to St. Olaf. I came out of the Army a very different person than when I went in. I had matured there. I had grown up. St. Olaf now represented a very different learning experience than when I had first enrolled nearly three years earlier. The schoolwork was no longer intimidating. I enjoyed my classes and had more confidence in my ability to absorb and comprehend new and challenging subject matter. Most significantly, I met the girl I would soon marry, Elizabeth BettyJo Huizenga, or simply BJ to those who knew her.

We married immediately upon graduation and honeymooned in Boulder, Colorado. We loved everything about Boulder: The mountain vistas were spectacular, and the air at a mile above sea level was clear and crisp. The local high school hired BJ to teach home economics, and I began what I believed would be a long and lucrative career with the Equitable Life Insurance Company of New York. I was hired as an agent to sell a new life insurance product that combined the benefits of whole-life insurance with a low-interest, home-loan feature. It was a solid insurance product, and I earned $25,000 selling the package my first year, which, incidentally, would be equivalent to about $250,000 today. That was, of course, huge money back then, especially for a young man right out of school. BJ and I were on top of the world, doing well, loving the surroundings we woke up to every morning, and looking forward to each day. It almost seemed too good to be true. And, as it turned out, it was. All of that abruptly ended with a November phone call from Berwyn, Illinois, informing us that BJ's father, Peter Huizenga, had suddenly died of a heart attack, as had his brothers before him. Our lives shifted course.

November can be a dreary time in the Windy City—much more "early winter" than "late fall," and a time when sunshine and blue skies can be pretty scarce. I was eager to be as supportive as possible to BJ and her grieving mother when we arrived in Chicago.

The Huizenga family was faced with an immediate and urgent problem. BJ's mother and three widowed aunts were now the primary owners of a family business that had lost its management. Each of the Huizenga brothers had succumbed to early coronary disease. With a dozen trucks and a relatively small crew of strong, hardworking men, the family business, Ace Scavenger Service, was generating about $750,000 in revenue, about $7.5 million in today's dollars. My first thought was that selling the business would be the best way to provide income for BJ's mother and the widows of the other deceased Huizenga brothers. That would have been the end of it, except for one twist in the story. One non-family managing partner, Lawry Groot, took me aside to urge that continuing to operate the business would be a far better alternative than selling it. All Ace needed, Lawry believed, was an intelligent manager at the helm. "You should be that manager," he said. "I can teach you everything you need to know about managing this business. It's not that difficult."

By Christmas, BJ and I had returned to Boulder, resigned from the jobs we loved, packed up everything we owned, and relocated to Berwyn, where we moved in with BJ's mother. We both hated leaving Boulder, but we had strong feelings that family always came first, and it was clear that BJ's family needed us to be there. And, truth be told, I was enthused about being at the helm of the business, especially with someone like Lawry to show me the ropes.

I rode every route with our drivers and quickly determined that Ace was well run. Our people took real pride in their work, and each driver considered every customer to be "his customer." They were hardworking, dependable men who were determined to arrive at every account on schedule and complete their routes each day knowing they had not given their customers anything to complain about. I liked what I saw, and it seemed my transition into management would be a smooth and seamless task.

But then, without warning, Lawry died suddenly early in the new year. For better or worse, Ace Scavenger Service was either going to thrive or not with me at the helm. With Lawry's passing, his brother, John Groot, stepped into his role, looking after his family's interest in Ace. He was as good a partner as I could have wanted. He supported every decision I made.

THE EARLY YEARS

Ace had twelve trucks and a terrific team of hardworking, loyal employees, and was servicing a few commercial and industrial routes. Ace's employees could certainly teach me more about serving our customers than I could teach them. BJ's dad had managed an efficient operation, providing reliable service to the businesses that relied on the company to be there on schedule, rain or shine, day after day, season after season. There wasn't much I could do to improve upon that. The greatest contribution I felt I could bring to the business was to grow it. This was, after all, the only business I could think of that provided a service that every home and business in the country that produced any waste at all was *required* to use. The utilization of licensed waste contractors was actually mandated throughout the country. The Huizenga brothers, who had operated the business for many years, knew how to serve their customers, but they had little interest in adding new customers. The Dutchmen who populated the industry in the Chicago area understood, more or less, that growing their business would mean diminishing another Dutchman's livelihood. No one was very interested in doing that. Routes simply weren't taken from another Dutch competitor, but routes were sometimes purchased and traded.

The business fascinated me. I knew that providing a very basic, mandated service represented an enormous opportunity, and I was certain that opportunities to substantially grow the business would materialize. The following twelve years confirmed the growth potential that I had anticipated, and I devoted the lion's share of my time during those years to transforming the small family business into a substantial, growing, state-of-the-art enterprise.

I had been at the helm of Ace for about a year and a half when I purchased, with notes to be paid over several years, three routes from another Dutch trash hauler, Herman Mulder, who had been a friend of my late father-in-law. I didn't give the purchase much thought at the time, other than it was an excellent opportunity to grow the business by acquiring additional nearby routes that we could efficiently service. That said, if one were to pinpoint a pivotal event that changed the course of the business I was running, and thereby the course of my business career, that simple purchase of three trash hauling routes would loom pretty large.

A few months after purchasing the Mulder routes, another trash hauler, Harold Vandermolen, from one of the larger trash collection companies in the Chicago area, contacted me. He had recently purchased a business, Acme Disposal, in Milwaukee. He proposed trading that Milwaukee business and a couple of routes in Wheeling, Illinois, for the three Mulder routes I had just purchased. He saw an opportunity to grow the Vandermolen business in the Chicago area. I saw an opportunity to grow in an entirely new major market. The idea of expanding into Milwaukee and developing a disposal business and industrial, containerized business was, to me, the most exciting development since assuming the management of Ace. That transaction, my acquisition of the business in Milwaukee, was the game-changer that focused my attention on opportunities far beyond the suburbs of Chicago.

The move into Milwaukee required recruiting operations and marketing talent to turn that acquisition into a prototype or springboard for expansion elsewhere. I first hired Stan Ruminski to manage Acme. He later hired Don Price to lead operations. They were a superb team, and Acme quickly became a textbook model of a well-run, full-service waste-handling business. Both men spent the balance of their careers with me, playing key roles again and again as we addressed ongoing opportunities throughout the United States and abroad. They were the earliest additions to a management team that would soon become the envy of the industry.

There were other developments taking shape by the end of that first decade. The challenges of managing the ever-growing volume of the nation's waste had begun to attract the attention of government regulators and elected officials at all levels, including the federal government. The Solid Waste Disposal Act of 1965, the nation's first federal initiative involving municipal waste management, caught my attention. I was certain that the industry had to get ahead of the legislative and regulatory changes that were coming or it would be swept aside by a surge of government initiatives. Ours was a highly fragmented and provincial industry. No one was really interested in, or paid any attention to, what was happening beyond the community in which they operated. Our waste industry had no presence in Washington, nor did we have any capability to anticipate, let alone influence, what changes legislators and regulators might impose on our industry.

For me, government regulation and legislation were uncharted and unfamiliar territory. I just knew something was coming. The Dempster Company, the primary manufacturer of large commercial and industrial waste containers, had formed an industry group, the Detachable Container Association (DCA), as part of its trade relations program. I was asked by Dempster to become the first president of DCA. As president of DCA, I met other large waste contractors throughout the country, and I decided to contact some of them to discuss the need for our industry to have representation in Washington. No one knew anything about what was or wasn't happening, but I was able to convince several of the contractors to join an effort to learn what was percolating in Washington, DC. We researched the names of companies there that specialized in representing businesses or industries in the nation's capital. We selected one of those firms, Larry Hogan and Associates, to meet with us and explain how they might represent us.

The meeting took place in November 1966 at the Chicago Athletic Club. The decision we made at that meeting to fund a national association to represent our industry would have an immense impact on the future of the industry and a substantial effect on me as well. Two young partners ran the firm, the late Larry Hogan, an attorney, former FBI special agent, and, at the time, a top-notch public relations strategist (whose son has served as the governor of Maryland), and Hal Gershowitz, a twenty-eight-year-old marketing and communications professional. Fifty-six years later, Hal still remembers receiving that letter from Ace Scavenger Service inviting his firm to meet with us in Chicago.

Within a short time, Hal developed solid working relationships with the relevant legislators and congressional staff professionals who were just beginning to focus their attention on the issue of waste disposal. He also began meeting with executive-branch personnel within the Public

Health Service and with leaders of environmental organizations who were also formulating policy positions with significant implications for the management of the nation's waste.

It soon became clear to me that we needed a full-time presence in Washington, DC and I began conferring with others in the field to convince them of this need. I led an industry-wide effort to retain Hal as the full-time executive director of our industry's new trade association, the National Solid Wastes Management Association (NSWMA). This was an enormous step for the industry and for me, personally.

Hal had established a modest waste industry trade show that supported the industry's initial efforts in Washington, DC, and he believed there was a market for a much larger trade show that, if successful, could support an impressive full-time operation. Virtually everyone I called listened attentively and, ultimately, supported my proposal to set up operations in Washington. The trade show Hal had envisioned was as successful as he thought it might be and, in a short time, WASTE EXPO became the largest trade show of its type, attracting public officials and private sector businessmen, engineers, and technology specialists from throughout the United States and abroad. Hal hired a top-notch technical staff, directed by a young professional recruited from Booz Allen Hamilton, Eugene Wingerter, to work with regulators and plan conferences at which talent from the private sector could offer insight to technocrats from the public sector. We had effective full-time representation in the nation's capital for the first time. The National Solid Wastes Management Association (now the National Waste & Recycling Association) with Hal Gershowitz as its chief executive was a significant turning point for the industry, and working with Hal on industry initiatives was an important new dimension to my career.

Within a few short years, the private waste management sector, which had consisted of a few thousand small, insular contractors whose interests were rarely focused beyond the communities in which they operated, was transformed into an industry eager for information about new technology, emerging public policy, and the seminars, publications, and equipment expositions the Association was providing. For the first time, I was now dealing with issues and participating in decisions that far transcended the operations of my own company—valuable opportunities that I pursued enthusiastically.

Concurrently, another unrelated development was taking shape that would also have a major impact on the industry and my career: the early consolidation of the highly-fragmented waste industry. It began when two young, bright businessmen, Tom Fatjo, an accountant, and Lou Waters, an investment banker, purchased control of a publicly owned, NYSE-listed machinery company, Browning Ferris Machinery, and began buying small waste collection companies, making their new company, Browning Ferris Industries, or BFI, the first waste collection company to be publicly traded on the New York Stock Exchange. They immediately began acquiring waste hauling and disposal companies for stock and accounting for these acquired companies on the then-prevalent pooling-of-interests-basis, which allowed for the roll-up of acquired companies with their assets and liabilities incorporated into the parent company at book or historical value, with no accounting for goodwill, which accrued substantially to the earnings of the acquiring company.

Lou and Tom saw an industry comprised of thousands of mostly small, independent businesses that collectively managed approximately half of all the solid waste produced in the United States. They recognized both the need and the opportunity for industry consolidation. BFI was demonstrating that there was a robust capital market for public ownership of well-run waste collection and disposal companies. This was a dramatic new development that introduced enormous potential for financing growth on a scale never before available to the industry. Previously, all corporate growth in the waste industry beyond what could be funded out of current earnings was financed over time with debt. I was intrigued by what Lou and Tom were doing and with their vision of how the industry might evolve. I thought they were onto something.

FORMING WASTE MANAGEMENT

I formed Waste Management, Inc., in 1968 by combining Ace Scavenger Services with a couple of companies in the Chicago area owned by Larry Beck and Tom Tibstra. Larry joined our management team and focused considerable attention on landfill operations following our acquisition of a large parcel of land in the Calumet Industrial District on Chicago's South Side. Among our earliest and best hires was a young, decorated Vietnam veteran, Marine Captain Phil Rooney, whom I brought on board as my assistant to make sure everything I wanted to get done was getting done. Phil had actually worked at Ace while still in school, before enlisting in the Marines. He started on the ground floor, or, more appropriately, on the back of a rear-loading garbage truck. He had wonderful organizational and management skills and also enjoyed an uncommon aptitude for operations. He quickly became my operational right hand, ensuring the growing Waste Management fleet ran well and that the company's growing roster of customers was the best served in the industry.

While Milwaukee represented the first business we pursued outside of the Chicago area, we soon developed disposal operations in Indiana and Minnesota as well. Meanwhile, thirteen hundred miles south of Chicago, BJ's cousin, Wayne Huizenga, had built a substantial waste collection and disposal business in Broward County, Florida. Wayne was as intelligent, hardworking, and industrious as anyone I knew. He had begun in 1962 by buying one truck, servicing one route, and grew his business, known as Southern Sanitation, from that modest beginning. A few years later, when Wayne and his first wife divorced, I bought his former father-in-law's interest in the business and became his business partner. While I was able to help Wayne secure financing to fund Southern Sanitation's considerable growth, his business remained independent and was not, at that time, part of Waste Management.

From the time we consolidated our operations into the newly formed Waste Management, Inc., in 1968 until our first public offering in 1971, our focus was on building the best operations in the country and growing the company. We succeeded in expanding our business in

Illinois and Wisconsin, and late in 1970, we added Wayne's Southern Sanitation to the growing Waste Management portfolio of operating companies. We now were servicing more than 8,000 commercial and industrial customers and more than 30,000 residential customers who relied on us for top-notch service. Our fleet had grown from a dozen trucks to more than 120 collection vehicles, and we were providing over 8,000 containers to our commercial and industrial customers.

In 1970, Lou Waters and Tom Fatjo made a strong case for merging Waste Management with BFI, but ultimately, we declined. As vice-chair of Waste Management, Wayne was a superb partner and had no interest in merging our business with theirs. We were well along with our own plans to go public, which I was eager to pursue as long as Wayne was willing to commit to staying with the company for five years. He agreed, and we took Waste Management public the following year in 1971. Wayne was integral to the effort, and he stayed with the company for the next ten years. BJ's brother, Peter, a young attorney, also joined the company as a director and suggested that we retain Arthur Andersen as our certified public accountants. Arthur Andersen provided a wealth of young talented financial specialists who were essential to our preparations to go public.

Don Flynn, a brilliant CPA who was our lead Arthur Andersen auditor, schooled us in the advantages of pooling-of-interests accounting, which was what BFI was utilizing in its fast-moving industry roll-up effort. Don was among the first executives we recruited upon going public in 1971. We couldn't have found a more brilliant chief financial officer anywhere. No one was more critical to the company's success than Don. He not only recruited extraordinary financial talent that enabled us to absorb so much new business, but he was also very effective in schooling analysts regarding the financial potential of the business and the predictability of financial results inherent in a well-run waste company. No one in the industry understood the intricacies of finance better than Don, and once he was on board, we were always able to secure the financing required for any challenge we sought to pursue. The company's ability to attract hundreds of millions of dollars of debt and equity enabled us to pursue opportunities throughout the United States and abroad successfully.

Phil Rooney, like Don, was also indispensable to the integration of so many businesses into Waste Management. Given the rapid pace with which previously independent companies were now being acquired by Waste Management, Phil's skill and aptitude for transforming these former owner-operated businesses into smooth-running company divisions that were enthusiastically managed by their former owners cannot be overemphasized. It was an arduous task, and I don't believe anyone in the industry could have accomplished what Phil accomplished. He created a highly motivated team of managers from a cadre of very independent haulers who had, in the past, become successful doing things *their way*. He was the quintessential American success story. During his years with the company, he rose from the back of a garbage truck to the CEO's office when, many years later, I tapped him to become the company's chief executive officer upon my retirement.

In 1972, about a year after going public, it was apparent that we needed someone in senior management who understood the industry and who could devote the time required to represent our company before the myriad of government agencies and financial institutions that were now spending an enormous amount of time defining what their role was going to be concerning our industry. We also learned that BFI had been trying to recruit Hal for the same reason. So we decided that if Hal was going to join one company in the industry rather than represent the entire industry, we wanted him to be with Waste Management. Wayne and I offered Hal the company's presidency, which, happily, he accepted. He spent the balance of his career with the company, building a strong government affairs capability in Washington and across the country, representing the company before the investment community along with our Senior Vice President of Finance Don Flynn, and directing corporate domestic marketing efforts.

EXPANDING WASTE MANAGEMENT

If there was a perfect time to be in this business, it was then. The United States and the rest of the developed world were struggling with the challenges of adequately managing *the totality* of waste that their nations were generating. It became clear that the collection and proper disposal of household, commercial, and routine industrial waste was the proverbial tip of the iceberg. Diverse waste management challenges would require diverse waste management skills and diverse waste management technologies. And so, we established WMX Technologies as the holding company under which the various skills and technologies would reside to tackle diverse challenges on a scale never before attempted.

There was nuclear waste to be managed and plenty of non-nuclear hazardous waste. In the decade before Waste Management went public, in the 1960s, twenty-three nuclear power plants had been constructed in the United States, and triple that number would be built in the decade following our first public offering. Meanwhile, the US Environmental Protection Agency estimated that over 750,000 businesses were generating hazardous waste in the United States when we became a publicly owned company. More than 30,000 different sites were disposing of hazardous waste, some at the industrial property where the waste was generated, and the majority at an assortment of sites operating independently throughout the country.

Simultaneously, the severe health risk posed by widespread and often haphazard installation of asbestos-insulating material was also making the news, as was the ubiquitous presence of dangerous PCBs (polychlorinated biphenyls) throughout the country. Demand for new wastewater treatment facilities represented new opportunities for our newly public company. The 1973 oil crisis also created new opportunities to capture the energy value of waste through high-tech incineration and steam distribution, and we built electrical generating plants at the company's numerous landfills, employing methane-powered turbines. The company did it all and did it all superbly well.

It became quickly apparent that these waste management problems were crying out for attention in major cities on every continent throughout the world. No one had ever really attempted to address waste management issues beyond their own country's borders. I, however, firmly believed that as long as we could match the people with the right skills with the tasks required of each opportunity, Waste Management could apply its operations pretty much anywhere. We successfully responded to opportunities to manage wastes worldwide: in America, the Middle East, Europe, Asia, Australia, New Zealand, and South America. Later, WMX Technologies consisted of different specialized companies, each handling different waste streams. Each company had its own management structure, its own strategic plan, and its own responsibility for managing its growth and its risks.

THE PEOPLE BEHIND WASTE MANAGEMENT

The greatest satisfaction I experienced during my working career was that the company I was tapped to manage as a young man ultimately transformed hundreds of careers. I tried my best to pair talented colleagues with new and challenging tasks and did so with a remarkable degree of success. Rosemarie Nuzzo Parziale, among our first hires, was tasked with organizing a smooth running and cohesive corporate office, screening and essentially selecting new executive assistants for our officers and other key personnel. Rosemarie kept schedules, arranged travel, coordinated numerous meetings, and became the go-to office person for endless administrative tasks. She was remarkably efficient in increasing the efficiency of our rapidly growing management roster. Rosemarie, now Rosemarie Buntrock, still keeps me more organized and more productive than I would otherwise be.

During what I call our WMX Technologies era, we became the nation's leading provider of an ever-expanding array of environmental services. We consistently identified talent within the company's operating system to task with these new management responsibilities. In the process, individual careers blossomed in tandem with the company's growth. If I were to list the accomplishments from which I draw the greatest satisfaction today, it would be that I often saw the potential in the men and women I worked with beyond that which they saw in themselves. They invariably rose to the occasion.

The adage that *cream always rises to the top* proved true time and time again at Waste Management. I was always mindful of how circumstances propelled my career, and I was determined to provide new opportunities for our employees to grow as our company grew. I was simply "Dean" to virtually everyone at the company, from our personnel in the field to our executives at our regional and national offices. Our people were always eager to demonstrate that they were up to the tasks we gave them to manage. To this day, I marvel at how so many of our people responded to new opportunities, often opportunities and challenges few of us had any prior experience with. Perhaps it was simply sheer luck or, as I prefer to think, providence, that repeatedly provided me with both the opportunities and the talented men and women to pursue one challenge after another.

It would be folly to try to name those colleagues who contributed the most to our success. There were so many team members who consistently performed above and beyond what might be reasonably expected of anyone. My ever-expansive vision of what we might accomplish would have had no chance of succeeding without the incredible team we assembled as we grew. Every accomplishment recounted in this book is a tribute to the men and women who joined me on this journey.

I often think of those colleagues whose talent quite literally changed my life. Stan Ruminski and Don Price, two of my earliest hires, demonstrated with Acme, that first non-Chicago-area company we built in Milwaukee, that we could build smooth-running, profitable businesses anywhere opportunity presented itself and, therefore, convinced me that there was little that was beyond our capability. Phil Rooney and Wayne Huizenga were both remarkable operations executives without whom I could not have ever even contemplated trying to integrate as much new business as smoothly as we did. Finally, Hal Gershowitz and I worked closely together to guide the transformation of a multitude of disparate, small, independent companies into a cohesive industry of contractors that began to collectively address the challenges of managing the nation's wastes. In so doing, we created the framework for the consolidation of the industry that proved so essential to the evolution of national operating companies such as Waste Management.

My greatest disappointment at Waste Management was the decision to sell Waste Management International in order to concentrate on the company's basic domestic business. While we had successfully pursued opportunities to manage a wide range of wastes in the United States and abroad, there was a much broader spread in profit margins from one class of business to another and from one country to another. The more diverse the opportunities, the less predictable the margins. While the company today has retrenched back to the original business model with its more predictable profit margins, Waste Management had, in its day, mastered the tasks of managing all of the waste-related opportunities it pursued anywhere in the world.

Circumstances entirely unforeseen in 1956 propelled me, a young newlywed selling life insurance a year out of school, into the leadership of a new industry, managing the massive volumes of waste that nations all over the world discarded every day. As new laws were enacted, we recruited many of the nation's leading experts to guide the company in addressing each new challenge. In short order, the company would grow into the world's largest manager of municipal waste, hazardous waste, nuclear waste, recycled waste, and waste-to-energy facilities.

The events that unfolded after I was asked to manage a small waste handling company in a Chicago suburb placed me at the right place at the right time to play a leading role in the worldwide evolution of an entire industry.

This book is dedicated, with heartfelt thanks, to all of the men and women who made this journey possible.

—DEAN BUNTROCK

TEAM MEMBERS LISTED IN THIS CHAPTER

Steve Batchelor	John Morris
Samuel Caraballo	A. Maurice "Maury" Myers
Gary Crohan	Steve Neff
Chris Disbrow	Scott Purdom
Jim Fish	Dennis Ryan
Paul Grochowski	Glenda Schaller
Johnathan Haugens	David Steiner
John Karcz	Mike Toomey
Harry Lamberton	Jim Trevathan
James Loper	Mike Watson
Bob Marcione	Paula Zito-Baysinger
Linda McMahon	

1

WASTE MANAGEMENT IN 2021

Glenda Schaller awoke at 3:15 a.m. It is early May 2021, a workday. She needed no alarm. After thirty-two years on the job, her body is attuned to the hour she has to get up.

She can smell the coffee percolating in her kitchen; she set it to brew the night before. She showers and dresses. The fastidious Schaller has, as always, laid out her clothes the night before. Her uniform is high-visibility safety attire and is a luminescent lime. Her steel-toed work boots are pink, a comical concession to fashion in her line of work.

Schaller makes a salad for lunch, gathers her things, and stuffs them into her work bag. She grabs a travel cup of coffee and heads out the door to her car. She is on her way. It is pitch-black at this early hour. There are few cars and trucks on the darkened streets. The sun will not rise for another couple of hours. Within thirty minutes, she arrives at her destination on Pershing Road in Cicero, Illinois, the operational birthplace of Waste Management, Inc., the world's leading waste management company.

Schaller is a veteran driver for the company. She has driven every type of route during her career: residential, recycling, commercial, industrial rolloff, even container delivery vehicles. She has wheeled the big trucks carrying tons of materials more than a million miles. In past years, her one-hundred-and-twenty-pound body lifted many of those tons on different routes.

Today she is driving a modified rolloff truck—a "Mercedes," she says—equipped with a crane to hoist Bagsters into the truck's thirty-cubic-yard container box. The truck weighs about 4,000 pounds and rises eleven feet, three inches in height. The Bagsters, a Waste Management innovation, are plastic containers that homeowners can use to dispose of their less bulky household debris, more practical and less expensive than the big industrial boxes on construction sites.

As Schaller arrives, a parade of trucks assigned to Waste Management's earliest routes, the 4 a.m. starts, are already being launched from the brightly lit depot. One by one, the drivers stop briefly at the gate to check their backup lights. Schaller joins the other assembled drivers, about one hundred and twenty-five of them, for their shift. At 4:55 a.m., they head out to receive a manager's briefing on safety, route information, and other collection items—the "huddle," they call it.

Once briefed, the drivers march to their designated trucks for a safety inspection. They are deliberate; there is a procedure to follow, a list to be checked. Schaller performs her own inspection, manually checking tire pressure and doing a walkaround to eyeball the truck's condition. Schaller climbs into the truck's cab. Her first task is to sign into the vehicle's onboard computer, the OBU, she calls it. Schaller logs in to an app and, touching the screen, keys in her personal identification number and password, a timeclock that records her start, and initiates the system. The other drivers do the same on their trucks. They do not use a pen or paper. The trucks have all been upgraded in recent years; there is no longer even a radio to communicate with her route managers. If the need arises, she connects to her managers via computer.

The OBU displays a pre-route checklist to gauge the condition of the steel truck's anatomy: its oil, transmission, tires, and other mechanical items. Like a jet pilot running pre-flight checks, Schaller enters numbers into the OBU, a technology that has become second nature to her. If something were amiss, she would enter it on the screen.

The computer displays her set route for the day and the stops she will need to make. It is equipped with GPS and will record every stop, every collection, and issue a ticket for the system to charge every customer. It also has the capability to show a photograph of collection stops to help Schaller identify the collection location, though customers do not always place their waste where it is expected.

The truck is designed for safety. It is equipped with cameras inside the cab and outside, capturing views front and rear. It has alarms that signal when the vehicle's equipment has not been deployed properly. Safety is sacrosanct.

"There are so many alarms for safety," she says. "Letting you know your boom is not in position properly, so the alarm is going off, or the alarm is going off and you gotta figure out why it's going off. Is it because of the outriggers? Or is it because the boom is not in its proper spot? It's stuff like that. I don't remember alarms like that back in the day. I really don't, but everything is letting you know that something's not right."

Schaller welcomes the focus on safety that her district manager, Chris Disbrow, emphasizes in their daily briefings. The measures protect her, the public, the customer, and the company.

THE BODYBUILDING TRUCK DRIVER

Bob Marcione, another Cicero driver, is starting his day, too. He's been collecting garbage nearly all his adult life, except for the one year he left to become a policeman in a Chicago suburb. After forty-four years on the job, Marcione will retire at the end of 2021.

He began his career at Southwest Towns Disposal, one of the first companies acquired by Waste Management when it began its consolidation campaign in the early 1970s. He did yard work, painted containers, cleaned the yard, cut the grass, and even got doughnuts. The efforts earned him a job, first as a helper, then progressing to a driver. Back then, it was two men on a truck picking up fifty-five-gallon drums of waste.

Today, Marcione is a fanatic about health and a former bodybuilder. He starts his day at 2 a.m. with a routine of sit-ups and push-ups and planks followed by a breakfast of fruits and nuts. He calls it a light workout. He carries a photo of his younger self with ripped muscles. He is still in top shape.

Marcione is one of thirty-two commercial route drivers at the Cicero site and wheels a new front-loader, which he calls his "Cadillac." The truck is twenty-three feet long (six feet more with its bucket down) and thirteen feet, four inches high. It weighs 28,000 pounds and carries 7.5 tons of trash. It has power steering, automatic transmission, electric side mirrors, automatic windshield wipers, and air-conditioning. The front-loader also has cameras inside, on its sides, and in the front and rear. Its several fuel tanks are filled with natural gas. Nearly 50 percent of Waste Management's fleet—9,000 trucks—rely on alternative energy. Many display the words: "We run on clean-burning natural gas." The company has installed one hundred and forty-five natural gas stations to fuel them. It has also set a goal to have 50 percent of its alternative fuel vehicles run on renewable natural gas by 2025. The company is continuing to evolve from diesel emissions.

Marcione's workday normally starts at 4 a.m., and he will make eighty to ninety collection stops at apartment buildings, gas stations, restaurants, factories, tire dealers, and stores of every kind. His day, typically eight hours, ends only when the route is done.

Before he heads to his truck, he, too, participates in the huddle. "We talk about all the incidents that happened the previous day," he says. "Safety is the No. 1 priority to everybody here," he says. The drivers counsel one another on safe operations. For twenty years, the company's zero tolerance for unsafe behavior, known as "M2Z" or Mission to Zero, has ruled operations.

"We have safety training all the time, all the time," he says. "Like wearing the vest, properly securing the truck when you get out, making sure with everything in your power that what you're going to do is not going to hurt you or someone else." Safety is the talk of every meeting.

His truck, like Schaller's, is equipped with a computer and safety tools that were unimaginable to him when he first began. He reviews his stops for the day on the OBU. "That's what we usually do in the morning," he says. "We review our stops to see if anything is new. You'll get stops like construction offices that are seasonal; that's only there when they're building a building and sometimes they'll just pop up. You got to scroll down to all the stops at the end of the day." And like Schaller's truck, if his dispatcher needs to reach him, it's by computer.

"The computer does everything," he says. "It's a GPS; it's got all the route information on the computer. It's a telephone; it's the only way we communicate." It has the capability to follow the driver on his route throughout the day and snap a photo if a stop presents a problem. It can respond to a customer who complains of a missed pickup and provide a photo of the stop in real time with the precise moment the truck arrived. "If you get into an accident, God forbid, you could take pictures with that of the damages and everything," Marcione says.

His truck is outfitted with a front forklift that he operates with a toggle switch to pick up steel containers. He likens it to playing a video game. The toggle switch guides the forks to lift the container up over the cab, empty it, and put it back in place. A blade in the truck's body then compacts the waste. Marcione is watchful for tight quarters and overhead wires. His in-cab camera watches it all.

"I love my job," Marcione says. "This job bought me everything. You're basically your own boss. You go do the route, and nobody's really breathing on you or saying do this, do that. I mean, you got duties you got to do, but you go do your route, get the job done, and you go home."

CICERO'S DRIVER-TURNED-MANAGER

Disbrow, the Cicero division's senior district manager, recalls how times have changed since his younger days when he drove routes on the streets alongside Marcione. He remembers, facetiously, that his early training decades ago entailed being given a map. Eight years later, after some resistance, he was persuaded to join management. For the past thirty-one years, Disbrow, a native Chicago South Sider, has climbed the ranks and worked in several divisions in Illinois and Indiana before arriving in Cicero to run its operations. He is responsible for the site's business performance, problem-solving, and profitability.

He participates in monthly and quarterly reviews, which are constructive and conversational. They talk "challenges and opportunities," Disbrow says. The dialogue is about taking action. It is supportive, not Stalinesque, and no one at the Cicero site goes to the corporate gulag for dropping the performance ball. The culture is a positive one. "Let's face it, you've got to move the needle. You don't walk out of these meetings feeling demeaned. It's okay, we're having a discussion," he says.

The route managers for residential, commercial, and rolloff "make the site tick," Disbrow says. They are each responsible for twenty routes. "They manage the towns, they deal with mayors and the village managers. They deal with every driver, and they coordinate among themselves," he says.

The route managers handle customer issues, special requests, and special pickups. Their performance, which includes safety, efficiency, preventable incidents, customer complaints or concerns, and idle time, is measured. "If we're not moving, we are not making money," Disbrow says. "The bottom line is meeting their efficiency goals and doing it in a timely manner and taking unwanted minutes and hours out of the line of business. They have a long day."

Data helps Disbrow do his job. He is armed with an app on his cell phone that uploads the information he relies on after midnight, giving him a head start for the day. "Before I leave my house, I could sit in my driveway and tell you exactly what's going to happen today, or what happened yesterday, what we need to stress or look at today. The technology is incredible." He knows he needs to be at the site, too. "You've got to have boots on the ground, facial recognition."

Disbrow is assisted by Paul Grochowski, deputy operations manager; Samuel Caraballo, senior fleet manager; James Loper, fleet supervisor; Mike Toomey, an operations improvement manager; Gary Crohan, a district controller; and Linda McMahon, a pricing analyst. There are two residential route managers, Dennis Ryan and Scott Purdom; two commercial route managers, Paula Zito-Baysinger and Johnathan Haugens; and rolloff route manager John Karcz. The Cicero drivers reach nearly 800,000 customers a month.

The organization is reasonably flat. Disbrow's role is just four levels from the company's CEO. He reports to Harry Lamberton, an area vice president, who reports to Steve Batchelor, one of two senior vice presidents at the company's headquarters who direct its geographic regions. Batchelor reports to John Morris, chief operating officer, who reports to CEO Jim Fish.

Across the US and Canada this morning, thousands of other Waste Management drivers and district staff are starting the day in much the same way. The streets are where the work begins, but the company's operations are much broader. After the waste and recyclables are collected, they must be processed, managed, and disposed of. The work involves employees on the slopes of the landfills and in the gatehouses, along the conveyors in the recycling centers, and in the sales and administrative offices. They are in the trenches of Waste Management, the environmental services business and industry giant.

MANAGING THE NATION'S WASTE

At the new corporate offices at 800 Capitol Street in downtown Houston, the news has been very good in recent years. There is reason for some satisfaction. There is also a desire to avoid complacency.

The company's reputation for strong management and ethical business conduct is routinely recognized. In 2021, for the third year in a row, the company was on the list of *Fortune* magazine's World's Most Admired Companies, cited for innovation, people management, use of assets, social responsibility, management quality, financial soundness, and quality of its services. It is a good list to be on. The Ethisphere Institute, a company that rates corporate ethics standards, has repeatedly named it among the world's most ethical companies. These and others are references the company's employees take pride in.

Waste Management possesses the most well-known brand in its

industry, a name as synonymous to its trade as Xerox is to photocopying and Kleenex is to facial tissue. As of the writing of this book, its stock trades are at an all-time high. The company's very name communicates a critical service, and the work its people do makes it one of the most important companies in North America.

The company is also performing well. At the end of 2020, Waste Management operated the largest network of environmental assets anywhere. It included 268 landfills, five of which handle hazardous wastes; 348 transfer stations, or the interim stops where the local garbage truck drops its trash for shipment to distant disposal sites; and 103 material recovery facilities (MRFs), marvels of complex technology and automation that process the recyclables put at the curb to go to end markets. Altogether, the company's assets total more than $29 billion. Its 2020 revenues exceeded $15 billion. About 80 percent of its owners are institutions, many the repositories of individual investors.

Jim Fish is the company's president and chief executive. At fifty-eight, Fish is tall, boyishly handsome, and obviously bright. He is the only member of management sitting on its nine-member board, of whom three are women. It is an active, independent board.

Fish began his career at the KPMG accounting and consulting firm and later got to know the future CEO of Waste Management, Maury Myers, while working at America West Airlines. Myers held the top job there. During a walk across an America West courtyard, Fish shared his plans with Myers to leave the airline for graduate school. Myers encouraged him to do so but only if he went to a top business school. Myers penned a letter of recommendation, and Fish was soon earning his MBA at the University of Chicago. After Fish graduated, Myers contacted him and recruited him as a vice president of finance at the trucking company Yellow Corp., where Myers was then CEO. But shortly after Fish joined, Myers departed Yellow.

"I pretty quickly decided I didn't like that business very much," Fish recalled. "The only person I knew there was Maury. I'd been in a couple of meetings with Maury, and all of a sudden he leaves and goes to become CEO of Waste Management down in Houston. I called him about six months later and said, 'First of all, thanks for leaving me high and dry here at Yellow. Secondly, what do you have available down in Houston?'" Jobs were available, Fish recalls Myers telling him. "I'm trying to rebuild the corporate office after the acquisition here. Why don't you come down?" he said.

Fish joined Waste Management in 2001, rejoining with Myers, a turnaround expert who had come to lead the then-troubled company in December 1999. Fish had ambitions. Myers had a message for him, Fish recalled: "If you were to go work for Proctor and Gamble, you'd have to get marketing experience because that's a marketing company. And if you were going to go work for Goldman Sachs, you have to get finance experience because that's a finance company. If you come work for Waste Management, and you want to move up in the company, you need to get operating experience because it's an operating company."

Fish also impressed Jim Trevathan, a longtime company executive who began his career in the trenches of hazardous waste sales and retired as chief operating officer after serving decades in a variety of senior sales and operating positions, including once running the Eastern region in Philadelphia. Trevathan, a native Houstonian, speaks with a friendly and persuasive Texas accent. He has a talent for reassurance.

Fish "was all over me," he recalled about his protégé. "'I want to learn the business at the front-line level,'" he remembered Fish saying. Some in the company doubted a finance person could ever succeed in the operational work of running a business. Fish, unsure he would get an operating role, was about to leave Waste Management. In fact, he had turned in his two-week notice. David Steiner, who succeeded Myers as CEO in 2004, got involved. "Let me find a job for you," Fish recalled Steiner stepping in.

Trevathan's persuasion and Steiner's support prevailed. Fish was soon running operations in Rhode Island and southern Massachusetts. He managed a landfill, a recycling facility, and some hauling companies. He did well.

Trevathan offered Fish more advice: "You've got to find a way to make sure the frontline team respects you as much as the corporate office team. Can you win over drivers and route managers like I can tell you win over CEOs and VPs? You've got to be able to do both, to run a district and to grow that business, and he did a superb job."

Fish impressed other higher-ups, moving from Massachusetts to Pittsburgh and then to Philadelphia to run the Eastern region as a senior vice president. In 2012, he returned to Houston, becoming the company's chief financial officer and then, four years later, earning the CEO job, succeeding Steiner.

WASTE MANAGEMENT TODAY

In October 2020, Fish oversaw the $4.6-billion acquisition of Advanced Disposal, a disposal, collection, and recycling company. The industry's largest company got larger, expanding its already strong presence in the east. In keeping with Justice Department antitrust concerns, Waste Management offloaded some operations to another company, the Canada-based GFL Environmental, Inc. It netted three million more customers in sixteen states. And in 2019, Waste Management bought Petro Waste Environmental, giving it a position to serve more oil and gas producers in Texas. As in days past, the company continues to add more companies to widen its operations in other communities. Acquisitions have been in Waste Management's DNA since the company was founded fifty years earlier. The pursuit continues as the company's phones ring with haulers interested in being bought out.

Waste Management's assets set it apart. Its business model, conceived in the 1970s, is resilient. Anyone can start a collection company. But disposal—and the network of landfill sites that Waste Management possesses—are far more difficult to replicate. Landfills are expensive. The odds favor failure in developing new sites. "It can cost you up to $400 million to ultimately permit and build a landfill, if you could make it happen. And in today's world, that's hard to do," Fish said. Analysts estimate that Waste Management and its two major rivals, Republic Services and Waste Connections, share more than 50 percent of the nation's disposal capacity.

At a recent conference, Fish was asked how Waste Management can be considered a sustainable company when it is the biggest landfill owner in North America. Fish was direct: "How much of that trash do you think Waste Management creates? The answer is we create none of it." According to the US Environmental Protection Agency (EPA), close to 150 million tons of solid waste are landfilled each year. The company disposes of 115 million tons a year of trash in its facilities. "If there were another solution for disposing of trash, then we'd be all over it. But at this point, the best solutions are these very highly engineered and very safe and very environmentally friendly landfills."

Waste Management is also North America's largest collector and processor of recyclables. "We're looking at alternatives to that," Fish said. "How do we use low-value plastics? How do we change the equipment in these recycling plants, which really hadn't gone through any type of evolution?" In response, the company has created a heavily automated material recovery facility near Chicago that relies little on labor and the hand separation of recyclables of earlier plants.

Today the company's growth is steady, predictable. Which is not to say any less aggressive or purposeful than it used to be. There is strong emphasis on internal growth, or, in layman's talk, growing the business they've got now, understanding their customers better, and then guiding the company's marketing and selling to attract even more. The customer, as in any business, is all-important. The company is investing in technology to better understand its customers and what motivates them to use Waste Management. The data they're mining informs their marketing pitch to attract new customers and retain them for a longer time.

Mike Watson leads these efforts as a senior vice president and chief customer officer on the nine-member senior leadership team. He is fifty-one and a company veteran, having arrived in response to an ad after finishing an economics degree at Indiana University. His first job was doing inside sales at the old Garden City division near Chicago's O'Hare International Airport, which the company has long served. He rose through the ranks, making stops as a regional vice president of sales and marketing in the Northeast, Canada, and the Midwest, then up to a senior area manager role, and then down to Houston to his current job. Along the way, he earned an MBA from the University of Chicago, where he got his "financial quant" background, he says.

Watson, unsurprisingly, is into measuring things and deeply researching his customer. He is excited about the investment being made in "the digitalization of the customer experience," he says. The company is apace with the changing technological times. It is not your father's garbage company. There is, he says, more equal balance among operations and sales and marketing metrics than there once was. To manage, one must measure.

So Watson and his group are mining the data. Gone are the days when folklore about customer requirements ruled. "We've done exhaustive customer research," he says. The data is yielding actionable insights into customers, but he will give away no secrets. The financial results show that they are getting results though. Watson enthusiastically talks about the company's go-to-market strategy, brand platform, analytics, e-commerce channels, leveraging the company's facilities network, its activities around sales operations, and the relative merits of inside versus outside sales.

Watson says the company understands that the customers' needs are evolving environmentally as well. "It's a constant reinvention of under-standing what our customers want," he says. Sales are "need-based," he says. "It's not 'I really can't wait to use more waste services.'" Customers want a partner who knows where they are going environmentally, he says.

PRIORITIZING THE ENVIRONMENT

The company tracks its own environmental performance and produces an environmental sustainability report every two years to measure its progress. The 2020 report, recording data through the year 2019, was ninety-seven pages long. Its charts point to progress. The company's primary sustain-ability goal is to reduce greenhouse gas emissions, which it plans to tackle in a few ways. Waste Management set a goal to offset four times the green-house gas emissions its operations generate by 2038. It's also increasing its volume of recycling while aiming to lower the level of contamination in the recyclables it manages to 10 percent, a serious impediment to end market acceptance. The company continues to educate the public about best practices, too. Greasy pizza boxes are trash, not recyclables.

Waste Management's engineered and lined landfills are equipped and finely tuned to gather the methane gas that buried wastes generate to create energy. It has ninety-seven landfill gas-to-electricity facilities and twenty-seven landfill gas-to-fuel sites. It participates in and supports sustainability fora and has a broad environmental agenda.

The company's operations are, of course, highly regulated. The nation's environmental laws, many of which were enacted after the company's founding, demand compliance: the 1976 Resource Conservation and Recovery Act (RCRA); the 1980 Comprehensive Environmental Response, Compensation and Liability Act (CERCLA, or Superfund); the 1970 Clean Air Act; the 1972 Clean Water Act; and the 1970 Occupational Safety and Health Act, along with many more state and federal regulatory requirements. Legislation and regulation are the fuel that powered the company's growth, and they still do.

The company continues to evolve, doing the daily work of tending to trash but understanding that the natural, commercial, and social environments—and its customers, the nation's culture, and the government—are evolving, too. It works hard to invest in its environmental sustainability efforts.

"THE GREENEST SHOW ON GRASS"

One other notable area the company is investing in is sports marketing. For years, Waste Management had provided funds to local divisions to sponsor professional sports teams.

Fans could see the company's logo displayed in Major League Baseball's outfields, on National Hockey League's arena boards, and on National Football League's scoreboards. It collected trash from the stadiums. In 2003, at Trevathan's recommendation, the company stepped up to sponsor the hugely popular NASCAR auto racing. Its trucks carried bright NASCAR decals, and it endorsed drivers Matt Kenseth, Sterling Marlin, and Bill Lester. The programs connected the company with car racing fans who followed the action in the pits and, more important, signed purchase orders.

In 2008, after years of NASCAR sponsorship, CEO Steiner embraced a new branding opportunity: a multiyear sponsorship of the Waste Management Phoenix Open PGA golf tournament held every February leading up to the Super Bowl. The event attracted more than 700,000 fans in 2018, far surpassing any other PGA Tour tournament in North America—and far more raucous than most tournaments' quieter, politer standards. (The tournament organizers, The Thunderbirds, no longer issue attendance figures, preferring instead to emphasize the event's charitable aspects.)

The company uses the annual tournament to host customers and to feature its environmental bona fides. Waste Management calls the event the "Greenest Show on Grass," and, working closely with the philanthropic Thunderbirds, conducts a zero-waste program, diverting event-generated materials from disposal through recycling, compost, donation, or energy creation. For the past decade, the company has also hosted a Sustainability Forum during the event to foster thoughtful discussion among environmental leaders.

The Waste Management Phoenix Open resonates across wide and important demographics, from C-suites to households, said Steve Neff, a retired vice president charged with developing and leading its sports marketing program. The tournament generates colossal name recognition and huge TV ratings for Waste Management—it captured 3.7 million viewers on the final day of the NBC broadcast in 2021. As of late, the company has endorsed PGA Tour player Charley Hoffman. His bright shirts, often green, advertise a large WM logo.

In 2020, the tournament served as a "soft launch" for a branding change, marking the company's continued evolution. The name "Waste Management, Inc." is expected to gradually slide into "WM," the green and gold supergraphics so prominent on its trucks and signage and now so closely identified with the company. Its move is not unlike the earlier branding migrations of companies such as UPS and AT&T and reflects the company's forward-looking direction and its future drive to meet both environmental and business goals. The company wants prescience, to the extent it can grasp it, to pay off.

PEOPLE FIRST

As part of his sustainability goals, Fish has established his "People First" objective, a goal framed with commitments to inclusion and diversity, customers, employee safety, and environmental sustainability. As of 2021, Waste Management has 55,000 employees. Treating employees well will serve the stockholders well, too, Fish believes. Next-generation leadership will come from the inside. Diversity, a strategic initiative Fish is driving, will be important. "This has not been historically a real diverse business," he acknowledged.

During an economic time when finding and retaining talent and labor has become more difficult and competitive, the company is leveraging its stock plan and other wealth generators as a selling tool. Fish enthusiastically describes how the company's 401(k) and other programs are delivering considerable wealth to employees. Thanks to their company stockholdings, drivers up to senior managers have amassed perhaps unexpected wealth, with many drivers earning hundreds of thousands of dollars and senior managers in the millions.

In the years since he's joined the company, Fish appears to have taken Trevathan's earlier counsel to "make sure the frontline team respects you." During the COVID-19 pandemic, no employees were laid off. The company guaranteed forty hours of pay to full-time employees, regardless of COVID-19-related service decreases. Other pro-employee programs help contribute to loyalty, customer service, and ultimately, in Fish's view, to reward shareholders.

No one can predict what's ahead for a company. But a look at the half-century history of Waste Management, its people, and its many adventures may inform the future.

It all began with the labors of immigrants and a vision that powered its way forward on the energies of twentysomethings with no experience in building a major company.

This is the story of how Waste Management, the nation's largest waste services company, arrived at where it is today.

Let us begin.

TEAM MEMBERS LISTED IN THIS CHAPTER

Dean Buntrock

Lawry Groot

Clarence Huizenga

Harm Huizenga

Peter and Betty Huizenga

Siert and Bertha Huizenga

Tom and Jennie Huizenga

2

THE ARRIVAL OF DEAN BUNTROCK

Waste Management, Inc. almost didn't happen.

That it did is a testament to a range of circumstances—some personal, some governmental, some timing, and some just plain luck and pluck—that propelled it forward. And, of course, there were the key decisions made by a group of men and women who had smarts, youth, and eventually, the resources to chase the opportunities that lay ahead. Most important, they were unafraid of the challenges and the hard work it would take to build the largest waste management company in North America.

Across the country in the late 1800s and early 1900s, immigrant groups had provided the muscle and willingness to do many of the unsafe jobs the more established would run from. It was no different in handling the nation's garbage and the distasteful remains of a growing industrial world and crowding cities.

Many of the new arrivals had planted themselves in the nation's urban areas. They came because they had to, escaping poverty, hunger, and lack of opportunity. For so many of them, collecting and disposing of refuse would create their livelihoods and feed their families. The Italians took over garbage collecting in New York and in Boston, where they faced stiff competition from the Irish. The Armenians traveled west to California. The Germans had St. Louis, the Jewish immigrants had Washington, DC, and the Dutch had Chicago.

Many of the early garbagemen were farmers who collected food wastes to feed their animals. In the beginning of the twentieth century, waste not collected was burned. It was hard, hard work. While garbage collecting provided a living, it sadly saw many of the men die too young.

These immigrants were simply trying to raise their families and make a living. No one knew what was ahead. It was way too soon for any big plans to be formed or grand visions to be dreamed.

HUIZENGA & SONS HAULING

The Ace Scavenger Company, begun on Chicago's West Side, was one of the many thousands of small garbage companies in America in the 1950s and just one of the hundreds of modest family-owned operations hauling waste in and around the city of Chicago. Ace had originated as Huizenga & Sons, begun by Harm Huizenga, a Dutch immigrant who stood five feet, five inches tall and started hauling refuse around the 1893 Chicago World's Fair with a single horse and wagon for $1.25 a load. Over time, his customer list expanded, and he added more horses and wagons. He left America briefly for the Netherlands after his wife died only to return several years later with a new wife. He started his company once again.

Huizenga was not unlike other immigrants who hauled waste in Chicago then. "Their meager economy taught them to work hard and live frugally," his grandson, Peter Huizenga, a former vice president, secretary, and director at Waste Management, said in a 1987 speech. "Their Calvinistic beliefs instilled them with a dignity and respect for manual labor and a strong sense of independence, while their strange language forced them into occupations that did not require English communication. And their cultural identity provided them with a sense of community and an ability to discuss their common business interests among themselves. These people came to America to find a better economic life. They didn't want to work for Americans, they wouldn't work on Sundays, and they didn't want to lose their culture and faith in the American mainstream. The garbage business was an occupation waiting for them."

Harm's sons followed him into the garbage business. Each started their own company. Tom, Siert (called Sam), and Peter built on their father's business during the 1940s and '50s after Huizenga passed away. But the hard work and long hours were deadly. Siert died at age fifty-five and Tom at forty-four of heart attacks. And then, on Labor Day 1956, Pete, the youngest of the brothers, passed away at age forty-eight. Like his brothers before him, his heart gave out.

During this time, new technology helped ease the physical toll on garbage haulers. Peter Huizenga recalled that the 1950s were the "renaissance days of the Chicago garbagemen. They had finally emerged from the times of horse and wagons, an immigrant society, depression, war years, and backbreaking physical work," he said. "New truck bodies were being developed with loading devices eliminating the need to carry drums and cans on your back up a ladder. The interstate expressway system was being built, expanding the territory in which a business could operate and travel. New kinds of dumps were being developed in old clay pits."

A LEADER IS BORN

Peter left behind his widow, Betty, and five children: Elizabeth, Sue, Ginger, Peter, and John Charles (J.C.). Elizabeth, known as BJ (for Betty Jo), had recently finished college at St. Olaf in Northfield, Minnesota, and married Dean Buntrock, the man who would turn the family-owned business into the largest waste management company in North America.

Buntrock had met BJ at St. Olaf's and studied there for three years. He found the academics more challenging than he had anticipated. His early education in Columbia, South Dakota had been at the one-room St. John's Lutheran School from grades five through eight. The school had thirty students, his high school class just eight. He didn't feel prepared for college, and it was a struggle, he admitted.

His education was interrupted his junior year of college when his father became ill, and he returned home to help with his family's business, which would then be sold. He then served in the US Army, attending finance school at Ft. Benjamin Harrison. His service educated and matured him. He was determined to return to school to complete his education.

The year before her father's death, in 1955, BJ and Buntrock had left Minnesota for Colorado.

They first stopped in Greeley, where Buntrock had some friends who were teaching and where he thought he might enroll at the state college to clean up a few hours of academics and complete his St. Olaf degree. He also wanted to see the Rockies, which he fondly remembered from a family trip out West when he was a child. He didn't find the Rockies in Greeley. Instead, he found cattle country with large feeding yards around the town and unappealing odors from the herds. "We're not staying here," he told BJ after just one night and a dinner. "What do you mean we're not staying here?" she asked. "I can't see the mountains," he answered.

So the couple moved on to Boulder, and Buntrock finished up his studies at the University of Colorado. Buntrock found academics much easier after the Army. The couple settled comfortably into a garden apartment, and both soon found jobs.

A local manager from the Equitable Life Assurance Company in Boulder recruited Buntrock, and he learned the life insurance business through life-insured home re-financings. (The manager's mother had been Buntrock's mom's hairdresser back home in Aberdeen, South Dakota, which is how they connected.) Buntrock made a lucrative $25,000 that first year and learned that in life insurance, no one came to you looking to buy. He had to find the buyers and convince them of the sale—a lesson that served him well in the future. BJ, meanwhile, was recruited by a school superintendent to teach home economics at a local high school. They were happy and content. Money was not a problem. Buntrock began plotting out his future.

A year later, when BJ's dad, Pete, died unexpectedly, the couple returned to Chicago for the funeral. While there, family, friends, and business associates appealed to them to give up Colorado and come back to Berwyn, the Huizenga family home. Someone needed to run the Ace business.

A NEW ERA FOR ACE

The Ace partnership was now primarily owned by the three Huizenga widows: Jennie (Tom's widow), Bertha (Siert's widow), and Betty (Pete's widow). There were a few other fractional owners who had other garbage companies but were, for the most part, uninvolved in Ace. Siert's son, Clarence Huizenga, was reluctantly managing Ace at the time. But Clarence was convinced he was going to die at age forty and didn't want the job. Lawry Groot, an Ace partner who was busy running a separate hauling operation, urged the young Buntrock to consider joining the business. Groot was concerned about the widows' welfare and future earnings. The business was all they had.

"You've got to come back. It's your mother-in-law," Groot told Buntrock at Pete's funeral. Groot had nearly the same number of shares in Ace as Pete's family. His wife, Virginia, was a close friend of BJ's. A World War II pilot, Groot had been shot down and held for more than two years as a prisoner of war. He was a take-charge kind of guy and wanted somebody to step in. Buntrock was reluctant; he didn't know anything about running a garbage hauling business. He was thinking he might go to Denver and work for International Harvester, where he had an in with a family friend, before venturing off on his own.

Groot was supportive. "Don't worry, I can coach you," he told Buntrock. With Groot's support, Buntrock decided to join Ace.

Buntrock wasn't totally unprepared. He had studied for three years at St. Olaf and learned finance at Ft. Benjamin Harrison. He had worked for his father's business in Columbia, South Dakota. His father, Rudy, had sold International Harvester farm equipment, tractors and trucks, Standard Oil products, and other farm-related supplies. Rudy had been a successful businessman, mayor of Columbia, and a board member of its local bank. Every sale seemed to need financing. Here, Buntrock had learned the workings of loans, creditworthiness, and cash flows. Of one thing he was sure: "I would have my own business," he recalled.

Managing Ace was a risk he was willing to take. Buntrock was not averse to risk and never would be. It would be the first of many risks to come.

And his responsibility would soon widen considerably as fate and mortality played its hand once again.

TEAM MEMBERS LISTED IN THIS CHAPTER

John Bakker	Lawry Groot
Herman Bosman	Pete Huizenga
Dean Buntrock	Archie Manning
Clarence Dagma	Julius Newsom
Leroy Elmer	Stan Ruminski

3

DEAN BUNTROCK BECOMES
A GARBAGEMAN

In 1956, just four months after Buntrock joined Ace, Lawry Groot, the man who recruited him and said he would teach him everything he needed to know about the garbage business, died. Buntrock was left to run the hauling operation on his own. He was 26 and still had a lot to learn.

Buntrock began by focusing on the business. Almost before anyone knew he had joined Ace, he began riding the routes with drivers. He spent time with them and studied their ways, their labors, and their customers. He got to know their hard work and heavy lifting and how they navigated the streets to reach customers. His introduction to Chicago waste collection also gave him his first glimpse of the big muscular city and how it worked. He found the size of the city intimidating but the business quite simple. He was up close in the front seat of a big rumbling truck getting street-smart.

The business was all about service. And so Buntrock also talked to Ace's customers. A quick study, he immediately understood that his drivers, the men who lifted the heavy drums and carried away the waste, were Ace's link to its customers. Every driver, every route was a profit center. He recognized the value of making drivers responsible for their customers. And he understood that service was not optional—the garbage had to be picked up. It was nondiscretionary. Laws enforced it. After riding with drivers, Buntrock knew more than ever that Ace had a very impressive team of drivers who were experts in their customers' needs.

Ace's customers ran down Madison Street on the city's near West Side and east into the city. Ace's downtown routes reached north on Michigan Avenue, along Clark Street and Broadway, and stretched along Lake Michigan shoreline reaching as far north as the suburb of Evanston.

"I was fascinated with the business," he said. "There were a lot of customers along Madison Street that had been Pete's father's business. That was where the original Huizenga business was. You'd go collect $3 or whatever, and this elderly gentleman is sitting by the stove warming himself. He'd always quiz you a little bit and say, 'I knew Harm [BJ's grandfather].' The bottom line was the customers were very loyal, and it was really the driver of the truck who was the company."

He also learned how stable the business could be. In Buntrock's father's business, his customers were local farmers who bought tractors, trucks, farm equipment, fuel, oil, and hardware. "Then we all prayed for a good harvest that would allow them to pay their bills," Buntrock said. The garbage business was different. It didn't have the risk that other companies faced, Buntrock thought. It was not seasonal. Customers needed the service in good times and bad. It was almost depression-proof. "You didn't have to sell them the service," he said. "They had to have the service."

The holidays came along quickly that first year, and many drivers received Christmas envelopes filled with cash from their customers. It was another sign of the drivers' value and their customers' strong appreciation for them. During breaks at the many restaurant stops, there was always a sandwich, coffee, or sometimes a whole meal awaiting them.

THE LIFE OF A 1950S GARBAGEMAN
Picking up the garbage in Chicago—especially in the downtown Loop—was and remains a night or early-morning business. From its then-base of operations at 4730 West Harrison St. on the city's West Side, Ace dispatched twelve trucks that handled eight collection routes, separate

dumpster routes for collecting larger containers, and two trucks for one-shot cleanups. Maintaining the trucks was a necessity, and the drivers had the support of good mechanics.

At Ace, a foreman came in before dawn, opened the office, passed out that day's worksheet, and took care of everything that happened during the night. The city trucks left at four o'clock in the morning at that time since they were supposed to be out of the downtown Loop business area by 6 or 6:30 a.m.; Chicago's stricter ordinances for the business district banned truck traffic during the day.

The drivers and helpers had to be strong. Much of what they picked up was ash in heavy steel drums. For the first half of the 1900s, many city households, equipped with incinerators in the alley, and businesses burned their trash in boilers and furnaces to reduce its volume. Downtown buildings stored their trash in sub-basements often several floors below street level. The drivers would haul fifty-five-gallon drums of ash up flights of stairs or on elevators to the waiting truck. They'd lug the drums to the sides of their trucks and sling them over their backs. The trucks were outfitted with hoists that would lift the drums and worker up to open body tops. Trucks loaded from the rear were just being introduced at the time.

Ace's employees looked out for one another. Many drivers came in even when they were sick. If they were too sick to work, the other drivers would split up their route books and chip in to ensure their customers' stops were served. It all started in the morning before the foreman even arrived. They were a team looking out for one another.

One employee, well-remembered, was Leroy Elmer. He had the physical build for the work, and he was younger. Elmer came to Ace as a helper loading drums onto trucks. Before long, he had earned his way to become a driver with his own route. Within a few years, he became a foreman. Many years later, Elmer retired as the manager of the company's huge Goose Island transfer station on the Near North Side not far from downtown, a key piece of real estate and a key part of the company's valuable Chicago infrastructure.

Another driver was Archie Manning. He, too, was big and a hard worker. While riding on a route with Manning, Buntrock noticed that his helper was paid nearly the same rate as the driver but didn't do as much work. Always looking for cost savings and productivity, Buntrock offered Manning a deal: Do the route alone without a helper and be paid nearly double the weekly pretax rate of about one hundred and sixty dollars. Manning accepted his offer. While some of the fellow drivers may not have been happy with the new arrangement, Manning was just too big and strong to argue with.

In his 1993 history of the company, *Waste Management: An American Corporate Success Story*, author Timothy Jacobson describes the company's recruitment of more drivers. John Bakker, looking for work when his father was selling his garbage business, contacted his cousin Clarence Dagma, who was working at Ace. Following Dagma's advice, Bakker came in on a Saturday and joined Elmore on a truck the following Monday. "The first day, I'll never forget it," Jacobson reported Bakker saying. "All they said was, 'You follow us.' Ashes, ashes, and ashes."

There were other similar stories. The foreman, Herman Bosman, had heard about a man who had lost his job at a tire company. That man, Julius Newsom, was soon fixing flats in the shop and within a year was also on a truck hauling ashes, Jacobson wrote.

The drivers were members of the Teamsters Union, and everyone had a good relationship with the union's local chief, Larry Monahan. When needed, Ace could rely on the union's to send over quality workers. At Ace, they earned good wages and enjoyed solid benefits. Issues with the union didn't occur very often. For the most part, during the 1950s, Ace's relationship with the union and Monahan was strong, and issues with worker performance and behavior were few and far between.

GROWING ACE'S CASH FLOW

As the early years progressed, Ace needed access to money to sustain and grow its business. Pete Huizenga had developed relationships with a few suburban banks but hesitated to borrow from them. "He never used them because the Dutch didn't like to borrow money," Buntrock said. "They dealt in cash. But with all the widows and families living off the business, I needed to borrow money, have them [the widows] guarantee it, but not reduce any of the draw that they were used to." He was chiefly concerned about taking care of his mother-in-law, Elizabeth. The other partners, the Groots and Bilthuises, had their own separate hauling businesses and remained satisfied with what Ace generated for them.

Buntrock turned to local banks in the nearby suburbs of Berwyn, Cicero, and Oak Park for financing. Ace could buy its trucks with no payment down. A new truck and packer body might typically involve a $17,000 note financed over sixty months, Buntrock recalled. But the business needed more borrowings than the community banks could offer. The Oak Park bank even said it had reached its lending capacity for Ace and directed it to the downtown banks where lending limits were higher.

By the late 1950s, the truck makers and body companies began to offer their own financing packages to make sales of their equipment easier. In a few years, Buntrock set up a leasing company to purchase equipment for the company's growing businesses.

Dean's father-in-law had been ahead of his times. He had been looking for better ways to dispose of the waste they collected. He and some other haulers built the first privately-owned incinerator that converted waste to energy at 39th Street and Laramie Avenue, near the Hawthorne Race Course in Cicero just southwest of the city. The facility sold steam to its neighbor. Huizenga was the largest investor in Incinerator, Inc., which relied on guaranteed waste streams from Ace and others. But "Inc. Inc.," as it was called, although a sophisticated plant well ahead of its time, was a financial burden. To service its debt, it charged disposal rates that were much higher than those of landfills. Ace, its largest investor, was forced to pay a much higher disposal cost than others using much cheaper, poorly operated "dumps."

Under Buntrock, Ace was growing quickly, its revenues doubling every three years. In 1959, Buntrock turned his attention to a new opportunity: Milwaukee.

GROWING ACE'S ROUTES

Milwaukee was growing but still lagged behind Chicago in the post-war boom. Buntrock learned that Harold Vandermolen, a friendly Chicago operator, and competitor, had a few routes in Milwaukee that he wanted to unload. But Ace still didn't have the financial resources to just go out and buy a company, according to Jacobson.

Still, Buntrock found a way. He succeeded in a two-step deal. He first bought a few Chicago routes from another operator, Herman Mulder, and then traded them to Vandermolen to acquire his underperforming Milwaukee routes. In 1959, the acquired Milwaukee company became Acme Disposal.

Buntrock hired Stan Ruminski, a former sales manager, to run the Milwaukee operation. Ruminski was Buntrock's first hire outside of Ace. He had previously managed sales for container systems for waste hauling, and he had also worked for a janitorial services company. He was the best salesman in Chicago's waste business, Buntrock believed. The timing was right. Ruminski was a very hard worker, incessantly striving to have Acme operate more efficiently, tighten up routes, keep costs down, and increase productivity.

Initially, Ruminski operated out of his car, where he stored a typewriter and sales forms. After he'd successfully attracted a new customer, he'd go to his car and type out the contract. Deal done.

Ruminski focused on the fundamentals of the business, Jacobson wrote in his history of the company. He had an instinct for it. He knew what was needed to manage his operations better—things such as efficient routes, handling labor, tracking containers, ensuring maintenance, and controlling costs—and, just as important, looked ahead to exploit Milwaukee's growth and expand with it.

In 1967, Ruminski moved Acme's offices to Brookfield, a community west of Milwaukee. He was eyeing the future and wanted to participate in the city's growing spread. With the move came a chance to develop a landfill and make Acme's operation less reliant on others for disposal space. A landfill would also give Acme more control over its costs and destiny.

Ruminski learned of a widow who had little interest in operating a quarry run by her late husband. The two negotiated a ten-year lease, according to Jacobson, with Ruminski promising to fill the site. Acme now had its own landfill. After engineering the landfill to make it a model facility, the site was operated meticulously and covered daily with a layer of soil—and its reputation as well-managed would help the company develop other sites in the future. The landfill's upkeep also helped gain the approval of the community and the appreciation of local politicians, always a valuable consideration for garbagemen.

Acme's success demonstrated the value of having a good team. The expansion to Milwaukee created value. In addition to waste collection, it provided recycling and disposal capacity. And it brought a major contract in Milwaukee County. Acme, a beta test of sorts in expansion strategy, proved that Ace's leaders in Chicago could confidently delegate responsibility and rely on outside talent and energy. Along the way, Ace had put together the first elements of its future integrated waste services company: collection and disposal. Picking it up and putting it down. Landfills were prized assets. Acme became the model for future expansion elsewhere.

With Acme operating well, Ace's interests were growing, and Buntrock continued to scout for new opportunities. During this time, at the dawn of the 1960s, new national concerns about the environment began to emerge. The public's concerns for clean air, clean soil, and clean water gained traction. These concerns would have a massive impact on Ace.

TEAM MEMBERS LISTED IN THIS CHAPTER

Dean Buntrock	Harold Gershowitz

4

THE DAWN OF WASTE MANAGEMENT
ENVIRONMENTALISM

In the mid-1960s, a small group of waste company owners was enjoying an evening cruise off the coast of Fort Lauderdale, Florida. The night was sponsored by Dempster Brothers, a vendor intent on marketing the equipment the waste industry needed to contain, collect, and haul garbage. Dean Buntrock was aboard.

Also aboard that night were two young ensigns—most probably sanitary engineers—from the US Public Health Service. At that time, the Health Service was the federal department where garbage and the environment came together. The Health Service's long-running responsibility was to protect Americans from disease, particularly the kinds of pathogens that might come from improperly managed waste. And during the 1960s, before the Environmental Protection Agency (EPA) was created, the Public Health Service owned the federal government's responsibility to eliminate open garbage dumps and develop plans to replace them with more protective disposal facilities and waste management systems.

The Health Service was not inexperienced in the field. It had been studying the nation's waste problems for years. The ensigns were tasked with conceiving a plan to protect the nation's health from the growing volume of solid and hazardous wastes.

Enjoying drinks on the deck after dinner that night, Buntrock listened attentively as the two ensigns described the environmental challenges the nation faced and the ideas and steps they planned to recommend. He was struck by what he heard. Perhaps stunned. The ensigns, dressed in crisp Navy-like uniforms and shoulder epaulets, outlined moves they believed would revolutionize the nation's approach to solid waste management. They were considering imposing strict national and state systems to manage solid and hazardous wastes. More sobering, the ensigns were calling for the municipal government to oversee solid waste operations and operate them similar to public utilities.

Buntrock and the others may not have understood every meaning or consequence of what the ensigns foretold that night, but he recognized what it portended: The government was about to take charge of the manner in which waste was managed in a big way. There would be strict rules and regulations.

The requirements on local government and private garbage companies—previously left to operate on their own with little governmental oversight—would most assuredly threaten the very existence of the private industry. The old way of doing business was about to change.

THE AGE OF ECOLOGY

The 1960s ushered in a new era of environmental awareness. There had been the hard-earned boom into the 1950s following the war years. And now Americans were beginning to reckon with the quality and health of their air, land, and water. The country was concerned. In this new age, many believed that environmental governance needed to change. It soon would.

For years, privately owned waste industry companies had been going about their business collecting the refuse of their residential, commercial, and industrial customers and moving these wastes to their designated dump sites. These sites were not pretty. More often than not, they were just open spaces on land at the edge of town—areas removed from the urban centers. Some were in low-lying wetlands. The haulers would collect their loads and ferry them to these dumping grounds. There was little if any oversight.

Typically, the driver of the last truck into the dump site would strike the match to light the piles, leaving behind smoldering mounds of ash and emitting gases into the sky and surrounding neighborhoods. Previously, burning had been the principal means of reducing the ever-growing volume of materials. Customers would burn trash at their own facilities and haulers at the dump sites. The sites burned packaging, food waste, cans, bottles, tires, chemicals, and other output—leftovers that the nation's newfound consumerism and buying habits had generated. The dumps not only generated unsafe gases but attracted rodents, generated fires, and polluted groundwater.

The Health Service knew well the problems the nation faced. It had worked alongside the American Public Works Association to come up with schemes for better disposal sites and waste handling systems—tasks that the two ensigns were involved in.

In 1965, Congress passed the Solid Waste Disposal Act, the nation's first law aimed at improving waste disposal technology. The law encouraged efforts recommended by the Health Service and Public Works Association to improve landfill siting, construction, and operating practices. The Disposal Act, while well-intentioned, lacked specificity and had no teeth. As the public's concern for the environment continued to grow, more calls for tougher waste rules with bite and enforcement would follow.

As the Health Service continued its work, it mostly offered recommendations and engaged in education and training programs targeted at regulators, landfill engineers, and operators to close old landfills and develop sanitary sites. The agency issued publications for the public and for public officials. It followed developments among municipal waste managers and private companies.

By the mid-1960s, 94 percent of the nation's 17,000 disposal sites were open dumps, according to a survey undertaken by forty-one state agencies in cooperation with the federal government. Only 6 percent of the sites met minimal sanitary landfill requirements. According to the survey, "The other 94 percent failed to meet one or more of the three most important factors characterizing a sanitary landfill—no burning, no water pollution, and daily earth cover." The survey prompted the government to launch Mission 5000, a plan with broad support from service, trade, health, and environmental organizations, to eliminate 5,000 open dumps and burning dumps.

There was no national EPA yet. Its creation would be prompted by the new era of ecological concern and outrage. The coming regulations and laws would have massive consequences for the waste industry.

ACE GOES TO WASHINGTON

Rather than fight the coming battle, as many companies faced with increased regulation would, Buntrock welcomed it. Or, at the very least, he recognized an opportunity to get in front of the regulations, help shape them, and participate in what was surely coming Ace's way.

Since its early days, Ace had been a prominent participant in Chicago-area industry groups. There had been the Chicago & Suburban Scavengers Association which begot the Chicago & Suburban Refuse Disposal Association. These groups were a means for haulers to have their voices heard, to influence local rulemaking, and to jointly respond to any issues that would arise and potentially affect their businesses.

At the federal level, Buntrock had taken a lead role in founding the National Council of Refuse Disposal Trade Associations in 1962 to represent private waste management companies nationwide. Buntrock, Harold Vandermolen (from whom Buntrock bought his first Milwaukee routes), and Marshall Rabins, an operator from California, each contributed $5,000 to hire a lawyer to incorporate their modest group. The attorney had done legal work for the Chicago association. They understood they needed to be represented in Washington, DC, where all the environmental planning was starting to emerge. They required a more effective way—a louder voice—to engage with Congress and the Washington bureaucracy. The job required help and sophistication and smart people who could take on their cause and make their case.

In 1966, the three found Larry Hogan Associates, a small Washington public affairs firm. The company was run by a young pair, Harold Gershowitz and Larry Hogan, who were invited by letter to come out to Chicago to make a presentation. The two were somewhat reluctant to make the trip, even though they were eager for new business and curious. They drew straws, with the loser going to Chicago. The twenty-eight-year-old Gershowitz was the dubious holder of the short straw and soon found himself at the downtown Chicago Athletic Association, a businessperson's dining and recreational club on Michigan Avenue where many a deal had been sealed. It was there that he met Buntrock, pitched his firm, and won the account for $10,000 a year. The persuasive Gershowitz became the executive director for the trade group, which, two years later, in 1968, adopted the name the National Solid Wastes Management Association (NSWMA).

The garbagemen had now strengthened their voice in the capital. The first item of business was to lobby against a highway beautification bill requiring landfills to be screened—which, with its added costs, was viewed by the association's members as impeding their disposal operations. But they were babes in the legislative woods. They were too late. Meeting with Chicago Congressman John C. Kluczynski, they quickly learned why they so sorely needed help in the nation's capital: The bill, the Highway Beautification Act of 1965, had already become law. Kluczynski, in fact, was its sponsor.

"We proposed a compromise that the industry would create standards that would be in conformance with the Highway Beautification Act," Gershowitz said. The association offered to support rules to engineer the landfills and incorporate liner systems and other features. "The regs we produced were never designed to make it easier for us to operate but to raise performance levels we knew we could meet," said Gershowitz. "If you didn't have the talent and resources, you couldn't meet them. As long as we could deploy other resources, talent, and money, it put us in a good competitive position."

Through this proposal, Gershowitz had signaled how the industry—and importantly, Ace—planned to behave. Rather than fight regulations at every turn, he had shown they'd invest in the process, a sort of "we're not fighting, we're joining" strategy. Going forward, the association continued to shape, influence, and contribute to the development of legislation and regulations. It was a counterintuitive approach to the often typical corporate gut reaction that regulation slows progress and costs money. They would embrace, participate, and master the process. And this strategy would profit them.

As the industry and Ace firmed up its Washington, DC presence, the movement toward greater regulation of waste continued, step by step. An alphabet soup of environmental laws and regulation was on its way in the coming years. This development signaled to the garbage companies that survival would require size, scale, and resources. And lots of capital.

Buntrock was looking to grow, setting Ace on a course for the expansion it needed. He knew where to find assistance. It was practically in the family.

TEAM MEMBERS LISTED IN THIS CHAPTER

Larry Beck	Peter Huizenga
Dean Buntrock	P.J. Huizenga
Don Flynn	Wayne Huizenga
Neil Emmons	Tom Tibstra

5

WAYNE CONVINCES BUNTROCK NOT TO MERGE BUT PURSUE

Wayne Huizenga was his name. Buntrock's cousin-in-law, he would help drive Waste Management's growth and acquisitions over the next decade. The grandson of Harm Huizenga, Wayne Huizenga arrived in Florida with his family while in his teens in the early 1950s. His father, Gerritt "Harry" Huizenga, was the younger half-brother of Siert, Tom, and Pete Huizenga. Harry had moved the family south, seeking his fortune in Florida real estate as a home builder.

In 1955, Huizenga finished high school at the Pine Crest School, where he was active on the baseball and football teams and won a contest to name the school's gridiron team, The Panthers. He then worked construction for a few years before attending Calvin College, a Christian Reformed Church institution in Grand Rapids, Michigan, that attracted generations of Dutch students. The energetic but less scholastically inclined Huizenga didn't last there long and soon enlisted in the US Army Reserve.

Two days after leaving the reserves, in 1960, he found opportunity and good fortune at the same time when a door opened in the Florida garbage business. It seemed to be his destiny. He met with his father's friend, Herman Mulder, the same man from whom Buntrock had bought Chicago routes and traded them for business in Milwaukee. Mulder had a waste collection business in Chicago, Miami, and Fort Lauderdale, and the latter's three collection routes in Pompano Beach were doing poorly. Fearing he was being cheated there, Mulder needed closer oversight and was hunting for new blood to manage the business. Huizenga was his man.

In his book, *Waste Management*, author Timothy Jacobson describes Mulder and the young Huizenga's chance encounter over lunch: "Wayne was two days out of the Army Reserve and planning to go to work with his father, Harry, again building houses in Ft. Lauderdale," he wrote. "But before lunch was over, Mulder had his new manager. The older man dismissed the young man's protests that he didn't know the first thing about the trash business with a brusque 'You don't need to.' Whatever he did need to know, he learned in the time-honored workingman's way: on the job. Wayne Huizenga never lost the knack."

Huizenga kept Mulder's trucks rolling nicely for two years, ensuring the customers were served, the trucks fueled and maintained, and the business better off than when he arrived. In 1962, at age twenty-five, Huizenga learned of an opportunity to buy a small hauling business in Fort Lauderdale. It was only a one-truck operation, but it was ownership. It was a start.

The competitive Huizenga became his own boss. And within a few months, the one-truck operation became two, three, and then four. By the end of 1962, he returned to Mulder to buy the three routes he had earlier overseen, which were then being run by his father, who had entered waste collection when his construction business slowed. Huizenga named his Fort Lauderdale hauling operation Southern Sanitation and his newly acquired Pompano Beach operation Southern Disposal.

As Florida's development boomed in the 1960s, so, too, did Huizenga's operations. He won new customers and secured new contracts as new developments sprung up. There were customers to be won, and win he did. Southern Sanitation, whose trucks carried signage facetiously offering free snow plowing and wedding catering, was adding routes.

At the same time, Buntrock was growing Ace's routes in the north. Huizenga and Buntrock didn't know each other that well, but the two would talk quite often about their common interests in the industry and running and growing their operations. In 1966, an unlikely event would set the stage for their union: a divorce.

NEW PARTNERS

In his history of the company, Jacobson explains the origins of the union: Huizenga and his wife, Joyce, divorced. Joyce's father, John VanderWagon, had a stake in Huizenga's operations and wanted out after the couple's divorce. VanderWagon was concerned he was no longer involved in the enterprise and worried about the growing debt Huizenga was accumulating to buy equipment and support his expansion.

Enter Peter Huizenga, a newly minted lawyer, and Wayne Huizenga's cousin and Buntrock's brother-in-law. Peter proposed that Buntrock buy out VanderWagon's interest in 1966. The new union, supported by Buntrock's ability to borrow, gave Huizenga access to the capital he desperately needed to fund his expansion. Buntrock used the equipment leasing company he had established earlier to finance the growth of Acme and Ace to now fund Southern.

Buntrock was a welcome addition. Huizenga would run and build the growing Florida businesses, and Buntrock would make sure Huizenga had the financing he needed. In just a few years, Huizenga would join Buntrock in forming Waste Management.

But before that turning point, another partner joined the operation. His name was Larry Beck. Beck and his partner, Tom Tibstra, operated the Atlas hauling business on Chicago's South Side and in Harvey, Illinois, a suburb south of the city limits. Buntrock knew them through meetings of the Chicago-area scavenger association. At the time, Ace had 75 acres on the city's far South Side along the busy Calumet Expressway at 138th Street that linked industrial northern Indiana to Chicago. It was on the very edge of the city of Chicago, surrounded by the Little Calumet River, the tiny suburb of Burnham, and a forest preserve.

The property was called the Calumet Industrial District site, or CID. Buntrock had labored for several years to persuade the City of Chicago and its southern neighbor Calumet City to approve the permits both required for landfill operations. When they finally approved the permits in the mid-1960s, Ace had its own disposal site, which was the most important asset of the business. With CID, Ace could deposit its collected wastes and charge fees to other haulers to dispose of their loads.

Adjacent to CID lay a one-hundred-and-fifty-acre parcel of land, part of which was in the Calumet City suburb. Buntrock needed the additional acres for his disposal site, but he couldn't afford it alone.

He negotiated with the owners, absentee operators from St. Louis who hoped to develop the property for a chemical company. The price offered was too attractive for them to say no.

In 1967, Beck and Tibstra became partners in the purchase of CID. With Beck and Tibstra now involved and the CID acquisition complete, the group had a more substantial business. A few years later, Beck and Tibstra would merge their hauling business with Waste Management. With new business in Florida and Illinois, Buntrock was determined to organize the pieces he had assembled into a single holding company. In 1968, after twelve long years of hard work growing Ace—a third of his life, he thought—Buntrock incorporated the various businesses as Waste Management, Inc., a name derived from a trade publication that emphasized "solid waste management." The name was golden; it would become synonymous with the entire industry. In today's parlance, it had brand value. Buntrock had a vision, a plan. It would take three more years to execute.

Buntrock knew he had to think bigger to grow the company. The growth that was on his mind carried risks. He needed to secure capital. And the only route to the kind of capital needed was to take their company public. Huizenga was reluctant to lose control of his business, but he knew this path was necessary.

Buntrock began to set their plans in motion—though the public offering almost didn't happen.

AN ATTEMPTED TEXAN TAKEOVER

In 1969, a Texas-based garbage collector had gone public, the first hauler to be traded on the New York Stock Exchange. The collection business was founded by a twenty-eight-year-old Houston accountant who bought a single truck with intentions of expanding his operation. Thomas Fatjo, the accountant garbageman, and his partner, Norman Myers, teamed up with Louis A. Waters, a Harvard Business School grad who wanted to apply his degree and banking skills to their business. Fatjo and Waters knew each other from their undergraduate days at Rice University, and Waters had raised money for Fatjo through the investment bank he worked for at the time.

"Pretty quickly I decided there was opportunity there and joined him," Waters said. Together, they bought Browning-Ferris Machinery Corp., a New York Stock Exchange–listed company. They merged the two businesses, changed the name to Browning-Ferris Industries (BFI), and set off on a pioneering campaign to buy waste hauling operations across the country.

Like Buntrock, the BFI founders were looking ahead. By the arrival of the 1970s, cities were placing bans on generators burning municipal wastes, previously a common practice. (The federal government would prohibit the burning of waste at municipal landfills entirely in 1979.) The change provided a disposal business opportunity for waste companies. "All of a sudden, there was a great opportunity in the waste business on a national basis," Waters recalled. "Prior to that, there were just collection companies, few disposal companies.

"Our plan was to acquire collection companies, redo their fleets, and modernize them," Waters said. But their main interest was to acquire disposal operations, too. "These companies were readily available because they were in an emergency situation. Our plan was to do a national consolidation of the waste business," he said. "We were the first to do that and one of the first to do a national consolidation, period, of any industry."

In its inaugural year, 1969, BFI acquired two companies in Texas.

The next year, they acquired a dozen more, and in their third year, nearly fifty. They were rolling. And they had the newly formed Waste Management, Inc., in their acquisitive sights.

"We sought them out," Waters said. "We looked at the roster of all the waste companies in the country. By then we were looking to acquire them. We had acquired about twenty companies. We were starting with the little ones, then we were going after bigger ones. They were one of the biggest."

BFI was working hard to gobble up important regional assets, and Buntrock, Beck, and Huizenga wondered if a BFI acquisition would be right for them.

In March 1971, Buntrock asked Peter and Wayne Huizenga to visit BFI, according to Peter's son, P.J. Huizenga. "Why don't you go down and hear BFI's pitch?" Buntrock had said. BFI dispatched their private jet to pick them up in Chicago and whisk them down to Houston. A helicopter ferried them to the top of BFI's headquarters in the plush Fannin Tower. No one in the waste industry had a private jet or helicopter back then, P.J. said. The Huizenga cousins saw the owners' luxury cars on display in the garage. They entered BFI's handsome boardroom on the top floor of the building. They were feted to lunch in the building's private luncheon club. "They had caviar and shrimp and champagne," P.J. said. "They were wined and dined and told how great their lives would be down in Houston. [That] they'd be made men. Then they went back in the helicopter and back in the private jet."

Merging the two companies would have marked an industry milestone. Under the BFI proposal, Buntrock would become vice chairman and Wayne Huizenga, a man intent on being his own boss, vice president. The discussions had gone on for months, Waters recalled. BFI had done its due diligence. The deal was all worked out. The operating team had looked them over, done their analysis. The financial team and the Arthur Andersen & Co. accountants, who aided both sides, had done their reviews, and the lawyers had done their legal reviews. The Waste Management leaders were still on the fence.

The night before the deal was supposed to close, Huizenga traveled to Buntrock's home in Chicago's western suburbs to discuss it. They were meeting the BFI team the next morning at the downtown law offices of their attorney, Peer Pedersen, to close their deal. "Where Wayne was always somewhat reluctant to do the deal, you could tell the next morning that he didn't want to do the deal," Buntrock said. "I saw all the positive reasons for doing it, but still the negative was always wondering if we could have done the same thing or maybe even better on our own."

Buntrock made Huizenga an offer: If he promised to stay with the company for five years, they wouldn't do the deal. Buntrock knew that Huizenga wanted to be his own boss. And he was right: Huizenga agreed and stayed on for ten years.

The next morning, they went into the city and told Pedersen that the deal was off. Then they finished taking the formal steps to go public on their own.

WASTE MANAGEMENT GOES PUBLIC

Waste Management would go public only three months later, in June 1971, surprising their counterparts at BFI, who did not know they had been working to take the company public. Buntrock later said that no major actions were taken and no critical decisions made without both he and Huizenga agreeing on the plan first.

The march to go public required work. The company had to prepare a number of financial statements of the combined companies. Peter Huizenga was also a partner in the Hlustik, Huizenga & Williams law firm, and Waste Management's corporate secretary. During preparations for the public offering, Peter played the role of an orchestra conductor. He assembled a team of financiers, accountants, and lawyers. He engaged the Arthur Andersen & Co. accountants from the firm's small business unit to review their numbers and prepare the financial statements. Peter's neighbor in Southwest Michigan, Fred Steggerda, a Dutchman, was an Andersen partner and directed the Waste Management work.

Peter identified the underwriters and selected the securities lawyers needed to dot the i's and cross the t's to meet the requirements of the federal Securities and Exchange Commission and state regulators. He hired the investment bank Chicago Corporation to underwrite and prepare the offering. Neil Emmons, a partner in the investment firm, introduced by Beck's father-in-law, led the effort with help from associates Phil Clark and Robert Podesta. Emmons would later serve on Waste Management's board.

Chicago Corporation's legal counsel was the law firm Bell, Boyd, Lloyd, Haddad & Burns. It was represented by Bill Burns, a senior partner, and associates John Bitner and Jack McCarthy. Waste Management hired the prestigious—Buntrock would say "silk-stocking"—Mayer, Brown & Platt law firm as its legal counsel. Its attorneys Dave Michaelson and Harry McKee handled this work.

The offering moved forward. The offering prospectus included three years of consolidated financial statements ending on December 31, 1970, which was prepared by the Andersen accountants. The prospectus described the company's businesses and assets in Illinois, Indiana, Wisconsin, Minnesota, and Florida. Altogether, the young company employed 240 people and owned about 120 trucks, 8,000 containers, and 150 stationary compactors. It used three kinds of trucks: 75 rear-end loading packer trucks and front-end loading packer trucks and 45 detachable container trucks. Its 1970 revenue was $10.4 million.

The prospectus explained how the packer trucks were fitted with hoists to lift and empty waste containers into the trucks. It talked about the company's transfer stations in Wisconsin and Florida with their tractors, trailers, and compactors that were "used to receive solid waste and transfer it into larger vehicles whenever the sanitary landfill site is distant from the area of collection." And it described Waste Management's landfills in four states—Florida, Illinois, Minnesota, and Wisconsin—noting with some prescience that "in this country, most solid waste is deposited or burned in open dumps or is disposed of by other undesirable methods. Legislation and public awareness of the effects of these methods on the environment are creating a demand for more acceptable methods of waste disposal."

The prospectus further described the eight original Ace partners: Elizabeth, Jenny, and Bertha Huizenga; Jacob Bilthuis; John and Virginia Groot; John C. Groot, a trustee under the will of C. Groot; and Buntrock. These partners were issued shares in the new Waste Management, Inc., in exchange for their interests in the Ace Partnership. After concluding negotiations for an exchange of substantial shares with the original partners, the prospectus listed the number of shares issued to Buntrock (50,068), Wayne Huizenga (64,800), and Beck (65,912). The original partners maintained their sizeable stakes in Waste Management, and a small number of others who contributed in the early days were also awarded some shares.

Chicago Corp. launched Waste Management as a public company on June 17, 1971. It was joined by fifty-one co-underwriters, who together agreed to buy 120,000 shares. The offering was for 320,000 shares at $16 per share. They raised approximately $4 million.

After the offering, the stock price rose sharply, ascending to around thirty-two dollars a share. A big New York bank—J.P. Morgan—was buying up all the shares it could put its hands on. "They bought everything," Buntrock said. "I'm guessing a week after we went public, they bought over half the outstanding stock."

Waste Management's leadership enjoyed the moment. The stock's price had quickly doubled. They had money in the bank, so to speak. It was currency for future consolidation. They had good reason to be confident. And they would waste no time putting it to work.

At the time, Buntrock was sitting with Don Flynn, who had overseen the preparation of the Waste Management offering's financial statements. "'You know what the message is when you've got a thirty-two-dollar stock?'" Flynn asked Buntrock. "No," Buntrock recalled saying. Flynn said, "You gotta use it."

And so they did.

TEAM MEMBERS LISTED IN THIS CHAPTER

Debbie Adams	Ron Jericho
Carol Borg	Don McClenahan
Dean Buntrock	John Melk
Earl Eberlin	Brian Oetzel
Ben Essenburg	Rosemarie Nuzzo Parziale
Don Flynn	Bob Paul
Tom Frank	Don Price
John Groot	Phil Rooney
Harris W. Hudson	Stan Ruminski
Peter Huizenga	Jerry Seegers
Wayne Huizenga	Marv Veltcamp

6

THE ORIGINAL TEAM

Following the 1971 public offering, there was an immediate urge to speed up the company's growth. As promised in their public offering document, Buntrock saw profit and challenges in the country's concern for the environment. He shared in the country's sense of urgency to improve the environment. Buntrock and his executive team were also eager to acquire new companies and grow the company. They were in a hurry.

But first, the team had to widen its skill set. They needed to add talent and begin to organize a management structure before they could seek the good acquisitions—ones that were well-managed, added customers, added value. Ones that grew the company.

At this point, the number of key people involved in the corporate center was still small, but the group would grow quickly. They were working out of lower-level offices at 15 Spinning Wheel Road in Hinsdale, Illinois. The offices were anything but lavish—they were more accurately described as sparse and workable with metal desks and filing cabinets and partitioned work areas. Comfortable. Good enough for the start, but it wouldn't be long before the growth machine was energized, and they would require much larger, nicer quarters.

In 1968, Buntrock hired Ben Essenburg to manage operations at what was then Ace, allowing Buntrock to focus fully on the future of Waste Management. Buntrock intuitively understood the elements their business needed: strong human resource talent, operational know-how, and, above all, financial control and discipline. This would be the architecture of their structure. In the months leading up to the public offering and after, Buntrock recruited a handful of critical people, many of whom would shape the company in the decades ahead.

THE MARINE CORPS MANAGER

First among them was Phil Rooney, a manager fresh out of the Marine Corps who brought talent and organization to the group. He would ascend the company's ladder quickly.

Rooney had grown up as the son of a police captain in Cicero, Illinois. He attended the Catholic St. Philip High School on Chicago's West Side and then the Benedictine St. Bernard College in Cullman, Alabama, where he graduated in 1966 with highest honors.

After graduation, Rooney joined the Marine Corps, where he served as a second lieutenant and was deployed to Vietnam in 1967. While there, he was promoted to company commander of a transportation unit and led convoys of men and equipment in Danang between April 1967 and May 1968. He displayed exceptional leadership and organizational experience and would continue to do so throughout his career. By the time he left the Marine Corps in 1969, Rooney had been promoted to captain and had been awarded the Bronze Star and Navy Commendation Medal both with Combat V., the latter signifying heroism in battle.

He became Buntrock's first assistant in March 1969. He was not unfamiliar with the garbage business. He had been introduced to the waste industry when he was hired by John Groot, an Ace partner, to work summers at Incinerator, Inc. Rooney was bright, energetic, and did everything asked of him and more. Groot eventually passed him along to Buntrock.

Rooney, who was in his mid-twenties at the time, brought not only leadership and a keen talent for managing to Waste Management but also the enthusiasm of youth.

In the coming years, Rooney would grow with the company and rise to be its president and CEO in 1996. Buntrock trusted him and regarded him as his partner.

THE CONTROLLER

Next, Waste Management needed someone to organize its books and financial systems in preparation for its public offering. In 1970, Bob Paul joined as the company's first financial officer before its public offering. Paul was a 1960 graduate of Yale armed with a degree in industrial administration. His plans to go on to graduate school at Duke University were short-circuited when his father passed away that summer, and he returned home to Chicago. However, he had taken courses in accounting at Yale, which opened doors for him at the accounting firm Arthur Andersen. He took night classes at Northwestern and DePaul, earned his CPA credentials, and became an Andersen audit manager.

Paul had planned to work in Andersen's consulting business, but he left when he learned the switch from the audit staff to client consulting would mean a pay cut. He then joined Stanray Corp., a mini-conglomerate, for a time, after which he started to entertain offers from other companies. He even turned down an offer from *Playboy* magazine after his wife, sensing family controversy and certain opprobrium, said, "You're not going there. How are you going to tell your daughters you were going to go there?" He learned through his Andersen connections that the newly formed Waste Management was looking for a controller.

The current Waste Management controller, Marv Veltcamp, a Dutchman, wanted to go back to Grand Rapids, Michigan, where his family was from. The Ace business had used a small accounting firm; its CPA was Maurie O'Mahoney. But now the company would need more firepower and a larger staff. After interviewing with his soon-to-be seniors at Ace, Paul joined the company and set about the task of getting the company's financial operation in more mature order in preparation for the coming public offering.

Paul needed assistance. He hired Brian Oetzel, who had been

working with Stan Ruminski in Milwaukee. Oetzel had previously worked for Clark Oil and Refining, where he had spent much of his time on the road auditing the company's operations.

In 1970 and early 1971, there were just a handful of people at the Spinning Wheel headquarters: Buntrock, Rooney, Paul, Oetzel, Don McClenahan, a consultant who handled marketing, and Rosemarie Nuzzo Parziale. Rooney had hired Parziale to offer administrative support and bring organization and more formality to the team. Parziale, who earned the job after impressing Veltcamp with her knowledge of financial statements, supported the entire team but focused on Buntrock and Huizenga. When the two loaded her down with their work, Huizenga told her to concentrate on Buntrock's needs. Parziale lived nearby, and her days were as long as the executives'. As her often six-days-a-week job expanded, she recruited Debbie Adams to support Huizenga and Carol Borg to assist the others. She contributed no small amount of order to the very busy and ambitious group.

They would recruit more team members, particularly in finance and accounting.

THE FOOTBALLING FINANCIER

Waste Management's leaders had relied on Arthur Andersen to prepare the financial statements required to go public. Andersen's accountants spent weeks digging into the financial records of the former independent businesses in Florida, Illinois, and Wisconsin that now comprised Waste Management. It was then not yet a large company and, appropriately, the task was assigned to Andersen's small business division, which concentrated mostly on serving modest family-owned enterprises. This was not a comedown for Waste Management.

Perhaps more important, the Andersen connection would prove to be a fortuitous marriage as it led the company to identify and recruit much of its top financial talent, conforming with Buntrock's plan to have a financially-educated office manager in each of the operating companies. Andersen's people impressed Buntrock, and he valued that the firm operated its own training center in St. Charles, Illinois.

No other accounting firm was interviewed when Peter Huizenga recommended hiring Andersen. And while Waste Management's senior team thought that all the Andersen people were extremely competent, one person stood above the rest: its audit manager, Don Flynn. Flynn had advised and assisted the Waste Management team through its public offering and in its first months.

Flynn, a stocky former college footballer from a South Side Chicago Irish enclave, graduated from Marquette University's business school in Milwaukee. He joined Andersen when his former employer, the David Himmelblau & Company accounting firm, merged with Andersen in the mid-1960s.

While at Andersen, Flynn asked his wife-to-be, Beverly—who then worked in Andersen's graphics group—who was the best leader to work for and learn from. She pointed him to Jerry Seegers, a senior accountant who would also later join Waste Management. "We became very close at that point because he worked quite a bit for me," Seegers said. "Through the years, he worked on a lot of other good accounts, both mine and Fred Steggerda's, who was the partner in charge of Waste Management."

"Don was always looking ahead. From the time I knew him, he wanted to make money," said Seegers. "He wanted to live a good life, and that's no negative. It's a positive. Don thought he would run into companies he could eventually merge or have acquaintances with."

Waste Management's leaders wanted Flynn to head their team and build out their financial organization. Huizenga and Buntrock made Flynn a "very good offer" at a restaurant over a Chicago toll road on their way to the airport, Buntrock said.

Flynn asked his former mentor Seegers for his advice. "Flynn liked the people," Seegers said. "In major detail he told me about Dean Buntrock and Wayne Huizenga and what they were doing. He said 'I don't know whether I should go. I've talked to Fred Steggerda, and he said I should stay here and make partner.' I said, 'Don, it almost sounds to me like an opportunity is knocking that doesn't come very often. I'll tell you what I would do. If I were you, I'd take the job if you really like the people. If it doesn't work out, I'm strong enough at Arthur Andersen and you're doing a good enough job, and you can come back. I'll give you a safety net.'"

Wayne and Buntrock had just landed in Ohio when they heard that Andersen had closed down its audit of Waste Management. "You must have made someone an offer to compromise the audit," their then-finance director Paul said. Flynn had decided to join the company as vice president of finance.

With Flynn's arrival, Paul took on responsibilities for treasury and for developing banking relationships and securing credit lines.

Buntrock had taken the lead in managing the company's financial needs up to that point. "But I didn't have a clue what was ahead," he said. "Don brought all that. He was uniquely different and so astute." Flynn knew accounting, but it would become clear he knew business, too.

The executive team recognized there was talent and expertise they could recruit from Andersen. In the days ahead, more than half of the financial and accounting managers would join from Andersen.

Among those joining was Flynn's lieutenant Ron Jericho, who was already familiar with the company after having scoured its records leading up to the public offering. Jericho had been a member of Flynn's small business group enlisted early on to help with developing Waste Management's financial paperwork. Jericho's work involved digging deep into past records and developing new consolidated financial statements based on the history of the business's development and acquisitions in Illinois, Florida, and Wisconsin. "We reviewed all the transactions, classified them in the proper accounts, and in that manner were able to create some financial statements for the larger entity," Jericho recalled.

Flynn insisted on hiring Jericho. He saw him as someone who understood the numbers and who knew the rules backward and forward. Jericho's attention to detail complemented Flynn's big-picture perspective. Jericho took over accounting and financial, and he later led the integration of the operating divisions' books with the parent company.

Jericho sensed the company had growth potential and thought it had competent people. And he saw genius in his mentor Flynn.

After joining Waste Management in 1971, Flynn would soon become one of the most important members of its leadership team, serving as its vice president of finance and ultimately chief financial officer and a director.

THE MASTER SALESMAN

After the public offering, Waste Management's leaders set their sights on participating in and profiting from the industry's quickening consolidation. That meant finding and selling haulers on why joining up with them would be a good move. Buntrock and Huizenga had just the person in mind.

His name was John Melk. Like Flynn, Melk earned a Marquette University degree but only after paying some workmanlike dues. He did a tour in the Navy and spent time laboring in a Cutler-Hammer factory job in Milwaukee. But he had larger ambitions.

Melk was a master salesman. Indeed, he seemed born to do it. After finishing college, he had rapidly advanced to become the national sales manager of the Heil Company, the manufacturer of garbage trucks and other heavy equipment for the waste industry. That experience would lead to a similar position at GarWood, also a maker of garbage equipment.

Melk had been active in waste industry meetings and developments; it was how he met sales prospects and industry leaders. Melk knew everybody one needed to know in the world of garbage—particularly the customers who bought equipment. It would be his knowledge of the industry, its companies, and its leaders that made him an attractive candidate and connected him to Waste Management.

Melk had known Ace's team since the early 1960s, although he was never able to sell the company much equipment. He had, however, come to know Huizenga well, and the two became friends in Fort Lauderdale. Melk was not happy in his current role at GarWood. Huizenga and Buntrock knew it.

"They came to see me one day in Detroit where I was living. They flew in and said, 'Hey, we're going to put our companies together and go public,'" Melk said. Then they made their pitch. They were taking the company public and needed help growing it. "'We'd like you to be in charge of acquisitions,'" Melk recalls them saying, "which I had no experience with. I kind of said, 'Sounds great.'" Melk knew the value his industry customers placed on access to capital. And he knew his customers generally faced a lack of it. All the new containers and big equipment cost a lot of money, and nobody had enough of it. The market looked promising for acquisitions, he thought.

Five months later, in November 1971, Melk was on the team.

SOLIDIFYING FINANCE AND ACQUISITIONS STAFFING

In April 1970, a young man from Chicago's West Side came to the company's attention. His name was Don Price. Brian Oetzel and Price had been Army friends while working together at Fort Riley in Kansas, where they received new recruits and assessed their intellectual and mental acuity.

Price first became connected to Waste Management through Stan Ruminski, who was looking for help running his Milwaukee operations. Oetzel had recommended Price, who had been working for a Wisconsin tanker truck outfit. Price interviewed with Ruminski and, as Price recalled, "a guy named Buntrock" followed. Price was quickly put on the payroll. He commuted between Chicago, where he lived, and Milwaukee. After six months on the job, he was called back to the company's Chicago headquarters. It was only weeks before the offering. He had a new job description: shadow Phil Rooney, see how he works, and see what needs to be done. Price soaked it up. He put his learnings to work as he led team members in fanning out across the country to search for waste companies interested in consolidating.

The company now had the finance and acquisitions teams slightly better staffed. They installed a management services group to offer corporate support to the operating companies, which involved other functions such as equipment purchasing, risk management, and human resources, and which ensured that partner hauling companies were current with best operating practices.

Key among the early staff recruits was Tom Frank, who joined from Allied Van Lines on the day of the public offering in June 1971, another in a line of Buntrock's eager and able assistants. John Groot, whose daughter was Frank's secretary at Allied, had encouraged him to visit the company. "'You ought to go over and talk to them,'" Groot told Frank. "That's what I did. It was a good discussion. My demands were not very high."

Frank had experience working in a variety of management roles for larger, more mature companies, including the Tribune Company, publisher of the *Chicago Tribune, New York Daily News*, and a string of other newspapers, as well as owner of a series of television and radio stations.

Frank was given a number of assignments, chief among them managing the company's insurance needs. As the early acquisitions were being considered, Frank reviewed their coverages and worked closely with Bill Boelke, an outside broker.

Frank marveled at the company's plans. "The acquisition program was premier," he said. And it brought its challenges. "They had all kinds of different risks, and they moved so fast."

EXPANDING OPERATIONS IN FLORIDA

Meanwhile, in Florida, Wayne Huizenga had been busy adding talent to his team, too. Throughout the 1960s, Huizenga had continued to expand his Southern Sanitation businesses with acquisitions in South Florida. Through aggressive sales efforts, he added customers and widened his service offerings. For example, after he saw industrial-sized waste containers used up north, he introduced them to Florida and started to rent them to Broward-area construction sites.

With a bigger company, Huizenga needed a bigger staff. In 1964, Harris W. "Whit" Hudson joined his operation and learned the first rule of Huizenga's management approach: Serve the customer. Hudson had been dating Huizenga's sister, Bonnie, and had asked for a job. He began working side by side with Huizenga in operations, sales, and management. It was the beginning of a strong relationship, and it would last through the years well beyond their time together at Waste Management.

With the stock offering completed, Huizenga's responsibilities transcended his Florida operations. His role was national. He was now executive vice president of Waste Management, and he needed someone to run Southern Sanitation. He turned to Earl Eberlin, an assistant city manager for Hollywood, Florida. Eberlin had risen from a sewage plant operator to the town's second-highest-paid official, running its public works, utilities, and water treatment operations. Eberlin had never met Huizenga. But Huizenga had attended Hollywood's town meetings in the hopes of winning its garbage business, and he had been impressed by Eberlin, who had worked for the city for fourteen years.

Huizenga wanted him to lead Southern Sanitation. He invited Eberlin to lunch at Manero's restaurant and then another lunch. Days and weeks passed with no answer from Eberlin. Huizenga, growing impatient, finally demanded an answer. Would Eberlin join Southern, yes or no?

"It stunned me because I was content with what I was doing," Eberlin recalled. He wasn't sure whether it was wise to give up his secure municipal position and salary. But Huizenga was persuasive. "The more I talked to him, the more excited I got about what he was proposing," Eberlin said.

Eberlin wanted Huizenga's assurance that he would be running Southern Sanitation without interference from other Huizenga family members, and he got it. He made one demand: He wanted a one-year contract with a second year assured in case things didn't work out. Huizenga told him there were no employment safety nets, but after persistence from Eberlin, Huizenga granted the concession. Eberlin got his contract. The concession would prove unnecessary.

Eberlin showed up to work at Southern Sanitation on a Monday, and Huizenga was gone the very next day, now constantly on the road to headquarters in Chicago and across the country searching for acquisitions. Eberlin didn't know much about the garbage business, but he was familiar with the haulers who used Hollywood's incinerator. And he had Huizenga's confidence and trust. That was all that he needed. Huizenga gave him the freedom to operate as he saw fit, and they communicated often.

"We had plenty of competitors in the area," Eberlin remembered. "What I was trying to do was maintain what we had, to keep the cities we had. We had all these competitors trying to take them away from us. We never lost one." The reason, Eberlin believed, was the emphasis he and his team put on customer service. It had been Huizenga's mantra.

A few months after Eberlin joined, he met with Huizenga, who asked him if he owned any Waste Management stock. "I said I cannot afford to buy any," Eberlin remembered. Huizenga reached for his checkbook, wrote Eberlin a personal check for $10,000, and told him to buy the company stock. "'If all goes well, you can pay me back. If it doesn't, forget it,'" Eberlin recalls Huizenga telling him. Years later, Eberlin wrote Huizenga his own personal check to repay Huizenga the $10,000. Huizenga tore it up.

WASTE MANAGEMENT TAKES OFF

The lens was now widening. The smaller management group was growing quickly. The management team had watched BFI grow and followed its model. There were more companies to bring into the fold and more people to hire.

Buntrock was at the center of it all. His influence, expressed quietly but persistently, would drive every major decision in the company for the next quarter of a century. He would get his people to take on huge projects they never would have imagined confronting and challenges they would overcome. He gave them confidence. His new employees were mostly young and largely inexperienced, yet they would grow the company, achieving goals they themselves did not believe possible. And that all began with acquisitions. Things were about to get much more interesting.

TEAM MEMBERS LISTED IN THIS CHAPTER

Dean Buntrock	Wayne Huizenga
Don Flynn	John Melk
Hal Gershowitz	

7

IN HOT PURSUIT OF PARTNERS

The word "acquisitions" was never really the right term to describe what the young Waste Management leaders were then about. Today, the word can often convey a sense of strategic takeover, of big institutions hungrily gobbling up assets and overtaking the weaker and more vulnerable, draining companies of their wealth and talent.

That wasn't how Waste Management operated at all in its earliest days. They were now in the business of building a business, and that would require partnerships. In the beginning years, they found partners by recruiting companies whose owners shared their common desire to grow and participate in an enterprise that would make them more successful, more secure, and, in the process, wealthier. For Waste Management and these company owners, the goal was to become bigger and better than they could be on their own. Waste Management would not pursue acquisitions but rather combine people of common interests and goals.

Waste Management's leaders knew their business from the street asphalt and truck axles up. They knew their competitors and the issues they faced. Importantly, they knew the pride the people in their industry had in their hauling companies and the investments they had made over the course of their lives in toil and time to become successful.

The businesses they sought had their own stories and legacies, rich histories unto themselves. The Waste Management founders were, in fact, just like them and approached them as equals. Each was an entrepreneur, each had overseen and built businesses. They respected one another. And they could talk to them about their common problems and opportunities, their accomplishments in building their companies, and the often-challenging roads they had traveled to do so. There was commonality.

Importantly, the owners of the targeted companies were not naïve. They knew the score. They knew they were sought-after and desirable targets in the rush to consolidate. After all, BFI and SCA Services, another publicly traded consolidator, had both gone public around the same time and wanted these companies as much as Waste Management did.

Upon joining Waste Management, the company owners—they became partners, really—continued to operate their local companies. They became the hauling managers upon whom the new company would rely for talent, know-how, savvy, and contacts to further expand in their areas. Paid with stock, they, too, were investors in the young company's future. Both parties' interests were aligned.

THE MOVEMENT TOWARD MERGERS

In the early 1970s, there were thousands of garbage companies hauling waste, nearly all of them family-owned operations that had strong ties to their local communities. The owners of these companies knew their marketplaces, they knew how to manage, and they had, for the most part, healthy relationships with their customers.

In the Chicago metropolitan area alone, the number of these companies was beginning to decline as Waste Management was becoming a public company. In 1955, the directory of the Chicago and Suburban Ash and Scavenger Association listed three hundred and nine members, most of whom were owned by Dutch managers. The association divided itself into geographic areas: West Side, South Side, North Side, Far West Side, and South Central. Each was led by an officer group. They'd meet once a month at places like Rosie's restaurant in Stickney or even the Washington Park YMCA at 50th and Indiana Avenue.

The directory noted that the association was formed "to advance the interest of ash and scavenger team and motor truck owners in the City of Chicago and its suburbs; to afford a means of common action; to encourage a spirit of cooperation; and to promote conditions favorable to the efficient and economic operation of commercial motor vehicles in the City of Chicago and its suburbs."

By 1971, the group had changed its name to the Chicago & Suburban Refuse Disposal Association, dropping the reference to "ash" to reflect the times—and the banning of incinerating waste—and how the nature of what they hauled had changed. It now had one hundred thirty-four members.

Headed by an executive director, Albert T. Hoekstra, the Association also included a new statement in its directory that its purpose was "to maintain the highest standards of excellence in the collection and disposal of solid wastes—to inform the public through all channels of communication the important role private contractors and their employees perform for them—to share significant technical information about equipment developments and operations techniques—to oppose unwarranted government interference but to cooperate with government in achieving necessary and equitable regulation to better serve the public."

The members worked with one another, and many attended Dutch Christian Reformed Church services in Chicago together. They enjoyed the kind of cooperation government officials might have looked askance at but which had served them well for years.

These Chicago companies represented a microcosm of the nation, and those operating in other cities were not dissimilar. They worked together, they competed, and they tended to know one another. These haulers shared the same future: They were well aware of the new movement toward mergers already launched by BFI in 1969.

THE HUNT BEGINS

Buntrock knew the Chicago haulers well and over the years, had come to know many other operators across the country, too. In 1962, he banded together with other industry leaders to create an industry organization, the National Solid Wastes Management Association (NSWMA) (today the National Waste & Recycling Association), to help strengthen the industry's voice in Washington, DC. The association meetings provided opportunities to meet fellow waste haulers from across the country who also recognized the value of having representation in the nation's capital. The association's hiring of Hal Gershowitz in 1968 had helped usher in a more professional approach to the industry's governmental affairs.

The local and national associations generated more familiarity among the companies. They also gave Waste Management lists, a starting point, of companies across the land that they could now try to recruit as partners.

After the 1971 public offering, Buntrock and Huizenga were eager to begin acquiring. But because the company was in an SEC-designated "quiet period" after its offering, they were prevented from acquiring for several months. When that period expired, they acted. At first, Buntrock, Huizenga, and occasionally Flynn, vice president of finance, began calling on companies.

By now, Huizenga, vice president of Waste Management, was traveling away from his Fort Lauderdale home every week. He would head to the company offices in the western Chicago suburb of Hinsdale but mostly for quick stop-and-go trips. His work was mainly on the road, flying and driving to the next prospect's door, to visit, talk about the industry, and, of course, work his wonders of persuasion. They were just getting under way and beginning to attract companies to join them.

Buntrock and Huizenga hit the road almost around the clock. "First we bought this cheap old Lear 23 jet after the public offering that had been owned by Arnold Palmer," Buntrock recalled. "I took Wayne out, and we made the first calls on all the haulers I knew from putting together the National Solid Wastes Management Association." They visited for hours talking to prospects, their families, and their employees. Before the year ended, John Melk, considered "the marketing guy" in the field because of his success selling trucks to garbagemen and deep knowledge of the industry and its players, had also joined Waste Management in the hunt for acquisitions. He was traveling, too.

THE 1971 REPORT

In 1971, Waste Management's first annual report to its new stockholders generously reported that "a formalized marketing department was established late in 1971 to expand merger and acquisition activities, to plan penetration of new markets, and to help local management develop existing markets to maximum potential." That was a kind of corporate-speak that dressed up the acquisitions approach. In fact, at first Waste Management's acquisitions team was chasing after owners they personally knew to have solid reputations, good companies, and, ideally, a presence and base in a given region. And while hauling targets were good, disposal sites were where the real value was added. The teams used its lists, and when they didn't have names, they'd arrive in a town and search the Yellow Pages. The hunt was unceasing. And these hunters were competing against BFI, which they knew had a similar formula—in fact, they were copying it—and whose leaders they knew were just as motivated to find the best local and regional partners.

The 1971 stockholders' report also advised, "We centralized and carefully expanded our corporate management group to include what we feel to be the finest talent available in accounting, finance, management services, market research, operations management, insurance, and computer programming and systems." The executive team was thinking ahead about the corporate functions they would need in order to quickly integrate, monitor, and support the companies they were trying to attract.

The truth was that building an organization was all new to its founders. The team's senior members were in their thirties, and the scouting and corporate employees were mostly in their twenties. They had little corporate experience—not that that would have advantaged or disadvantaged this brash and collegial crowd—but they had energy to spare and goals to meet. They were eager.

"There were few people who had any experience at what they were doing," Buntrock said with no little understatement. "Everyone had to figure out what their job was and what they could do. There really was no focus on criticism. It was learning on the job and what needed to be done."

They did, however, have a sense of how to recruit new companies. Huizenga and those on his search-and-conquer team traveled during the week, moving rapidly to knock on doors. They would return to the cramped Spinning Wheel office and, very soon after the public offering, to the nicer and newer office space at 900 Jorie Boulevard in the suburb of Oak Brook, to review progress and prospects and the deals to chase and close. However, the traveling teams spent little time there.

THE AGE OF PROFESSIONALISM

Waste Management's team had more space to operate in its new, spacious, white-columned building in Oak Brook. Each department had its designated space, areas for finance, acquisitions, and operations. There were offices with windows for the more senior managers, ample workstations at the center, and a conference room. There were "hot desks" as one manager called them, where employees could sit, draft reports, and make calls, and, just as quickly, race out of town. The fancy lobby had a chandelier. But form followed function, and each department was organized with a sense of order, rationale, and workflow. The staff was busy with the labor of bringing in businesses. "Everyone worked eighty-hour weeks," Melk recalled.

The teams were backed by a group of dedicated administrative aides who helped to organize and keep track of the traveling executives. They served to manage the executives' workloads and complete their reports. They knew what was going on. The aides' workday often continued after 5 p.m. They stayed longer to complete their tasks. "The officers who were traveling were really efficient because of the assistance they had back in Oak Brook," a retired executive recalled.

The leadership team met Saturday mornings. It was a system they established to discuss and review the increasing number of merger prospects the company's acquisition teams identified and pursued. Buntrock and Huizenga set the priorities. They knew the companies that belonged to their local industry organizations. They knew fewer people in locations where the owners were less interested in buying equipment, and so Melk, with his equipment sales experience, would identify these prospects more easily.

There was reason for enthusiasm. The industry was entering "an age of professionalism," as Waste Management's first annual report noted. The typical American generated about a ton of solid waste a year on average. Waste volumes were projected to grow by 60 percent in the 1970s. The opportunity presented by greater volumes of waste was "clear," the annual report said. So business, when done right, was going to be good. They were certain.

There had been another helpful and important development, too, for both the country and the company. The nation and its leaders had begun to understand the serious health and environmental threats that unmanaged solid and hazardous waste could pose. If the industry was entering an era of professionalism, it was also launching headfirst into an era of regulation, the strictest rules ever focused on waste managers. Wisconsin's Senator Gaylord Nelson, recognizing the public's concern, created the first Earth Day on April 22, 1970. A little more than two months later, President Richard Nixon signed an executive order establishing the EPA on July 9, 1970. Only five months later, on December 2, 1970, the EPA began its operations.

Waste Management's leaders understood the significance of these moments. By the end of that first year, 1971—only six months after the stock offering and only one year after the formation of the EPA— the young company had more than twenty subsidiaries. Businesses in Indiana, Minnesota, and Ohio had been added to the company's operations in Illinois, Florida, and Wisconsin. Revenues had grown 42 percent to nearly $17 million. The company served 40,000 residential customers and 14,000 commercial accounts. They had 13,000 waste containers and compactors and 200 collection trucks.

And they were just getting started.

TEAM MEMBERS LISTED IN THIS CHAPTER

Dave Blomberg	Dave Jorgensen
Bob Brach	Dave Kopp
Dean Buntrock	John Melk
Timothy Casgar	Brian Oetzel
David C. Coleman	Bob Paul
Tom Collins	Don Price
Bill Debes	Mike Rogan
Don Flynn	Phil Rooney
Wayne Huizenga	John Slocum
Royal Johnson	Fred Weinert

8

DON FLYNN'S FIVE-YEAR PLAN

Don Flynn always had a plan in the early years. The vice president of finance would frequently—perhaps facetiously—refer to it as "The Five-Year Plan." Trouble was, not everyone knew there was a plan or, if they did, even what it was.

"Would you let me in on it and tell me what it is?" one of his team members once asked. Others—a few at the highest levels—would ask the same question. Flynn was direct, succinct in his answer: "Go out and get as many companies as you can." Buntrock and Huizenga shared this plan, too. Whatever the answer, the senior team all agreed that "going out and getting as many companies as you can" was a good plan. The only plan, in fact. And that's exactly what they set out to do.

Following the company's 1971 public offering, the acquisitions teams hurried to find and size up companies to join them. They followed a simple playbook: marketing, operations, finance. The first scouts in marketing would meet the owner, convey the future they foresaw, sell them on the partnership, make the proposal, and handle the negotiations. They engaged in the straightforward dialogue of garbagemen.

An operations team would follow, capturing a street-smart picture of the physical business—the number of trucks, the routes, the age of the equipment, the number of containers, and so forth. A financial team then concluded the process—Bob Paul and Brian Oetzel from the finance group in the earliest days—to review the company's books, measure the receivables, and take stock of the seller's assets. They'd seal the deal. One and counting. More on the way—many more.

While the acquisition managers focused their sights on new target companies to enlist, the executive team was busy bringing on additional talent to shore up the staff. Training, personnel, and industrial relations became the department of employee relations, with David C. Coleman, the son of a Teamsters organizer, as its director. He had the kind of experience and background they needed. Who better to talk to the unions than the son of a Teamster? The team was moving so quickly to find and close the friendly mergers that they needed more hands on deck to review manage, monitor, and integrate the incoming assemblage of businesses and assets.

1972: THE NEW FINANCIAL RECRUITS

Bill Debes was one of the new financial recruits. He was a man who stood out—in his financial talent and physicality. Debes was six feet nine, a former DePaul University basketball player, and he was among the first Flynn recruited to his team. Flynn knew Debes from his time at Arthur Andersen, where Debes had worked on the audit of SCA Services, another public rollup company that was also gathering waste hauling and disposal companies and other services, including vending machines (Coincidentally, SCA even had an office near Waste Management's in the Spinning Wheel Road complex.) Debes was familiar with the waste industry, and he respected and appreciated its profitability.

Flynn was trying to fill a slot to handle financial reviews of the companies they were after. He needed Debes to look at prospects' books and assess their potential. Moreover, he needed Debes to hire more auditors to assist him. After interviewing with the other senior managers in Hinsdale—a rite of passage that moved quickly and seemed to be more of a courteous formality—Debes went to work for Flynn. It was early in 1972. Arthur Andersen had been doing most of the financial reviews until then, but Flynn, frustrated with the firm's premium price and lagging pace of its fastidious reports, wanted to bring the work inside. He wanted to save money and was looking to accelerate the speed of acquisitions. Flynn believed Andersen's pokiness had caused the company to miss out on a few deals as the audit firm hadn't provided the completed financial reviews needed to share with Waste Management's board so offers could be made and mergers could move along.

Debes knew where to look to find auditing help, and he moved quickly. "I had to bring in three smart guys of pretty strong stature," he said. "I took one guy from [the auditing firm] Main Lafrentz; I took one guy from Arthur Andersen, and I grabbed a guy from Pricewaterhouse who I knew in high school. All those guys turned out to be long-term Waste Management employees." The three were Mike Rogan, Fred Weinert, and Tom Collins, respectively.

Rogan, from Main Lafrentz & Company, had earned his accounting degree at Indiana University, not far from his hometown of New Castle, east of Bloomington. He passed up a steel company's offer to join Waste Management as treasurer.

Weinert, a University of Dayton grad, had known Debes while working as an auditor at Andersen's Chicago Loop office at 69 West Washington St. When Debes called, he was working for a plane-leasing outfit. Weinert, who later would head Waste Management's international operations, brought with him an entrepreneurial acumen gained while a Dayton student. He had been in student government and ran the university's concert programs, bringing in groups such as Diana Ross and the Supremes and Simon & Garfunkel—ten acts a year in the school field house. "I picked up my experience promoting concerts, and it helped put me through school," he said. He earned $600 a concert. And Collins, an accomplished auditor recruited from PricewaterhouseCoopers, had been Debes's friend at Fenwick High School near Chicago and would go on to have a long career with the company as a controller.

Flynn wanted to develop internal staff, too. Paul was charged with recruiting the early auditors and younger financial staff. He visited a number of colleges in Northern Illinois: The University of Illinois Chicago, DePaul University, Northern Illinois University, and others. As he recruited, he followed Flynn's dictum to place extra value on candidates who had worked their way through school. It served as a leading indicator of how they would perform, he said. "That they had worked their way through school showed they were hard-working. They were working for their own education and paying for it."

The new recruits understood the company was embarking on a spree of adding hauling companies. Their job was to conduct the due diligence, run financial analyses of the target companies, and examine their balance sheets. Other new recruits quickly joined the financial organization, including John Slocum, recruited by Collins from PricewaterhouseCoopers. "In 1972, they were in the midst of buying companies," Slocum said, who was initially uninterested in the garbage industry. His friend Collins persisted with repeated sales pitches accompanied by beers and cheeseburgers and, finally, a meeting with Flynn. Slocum relented and signed on. "It was really busy," Slocum remembered. Bob Brach and Royal Johnson, both of whom were from Andersen, also joined. Dave Kopp, from upstate New York, was enticed to handle tax matters, which would soon be more complicated as Waste Management entered more and more states.

Rogan was impressed with his workmates. "We kind of took each other for granted, smart people working with smart people. Maybe I thought I was smarter than I was," Rogan said. "We were so young. *Why were you asking us to do this stuff?* I remembered thinking. Dean [Buntrock] would say we were too young to be scared. Regardless of the reasons why, these people were very insightful about what their capabilities were. The company would be swift in making changes."

THE LAW DEPARTMENT
When the time came to close the fast-increasing number of merger transactions, the company needed a full-time, in-house lawyer. Timothy

Casgar was that person, arriving as the company's first general counsel just after the public offering in June 1971.

Like the attorneys who had represented them in the public offering, Casgar was a quality lawyer and very smart, possessing impressive educational credentials: Harvard College and Stanford Law School. He had first come to the company's attention in Milwaukee, where he had represented the Acme division in Wisconsin, which had contributed a significant contract with the City of Milwaukee to Waste Management's growing business base. The three-year contract "to receive, transfer, transport and dispose of residential solid waste collected by the City" was among those referenced in the company's public offering documents.

Casgar commuted from Milwaukee at first but then moved his family down to Chicago's Western suburbs. Like the other team members, he traveled nonstop working on the negotiations for the growing number of contracts and closings. It was exhausting, he said. In addition to acquisition closings, he simultaneously handled contract and regulatory matters and the frequent litigation generated by lawsuits that accompanied the day-to-day operations of the hauling companies. He worked closely with Curt Everett, John Bitner, and Jack McCarthy, the outside attorneys at Bell, Boyd & Lloyd, who provided additional help with mergers and some other issues.

"The work environment was go-go," Casgar remembered. "It was very educational in terms of rollups. It was a packed bag, a suitcase in the office. You never knew when you'd be leaving." At the corporate office, there were long hours and a culture that demanded workdays extend well into most evenings.

Casgar worked closely with Buntrock and traveled often with Huizenga. He and Huizenga would spend days kibbitzing with and cajoling merger prospects. "Trying to convince these guys to join the parade was always a big problem. That's why Dean and Wayne, having been in the business a long time, were able to convince a lot of guys that it was a good idea. Most of them made out pretty well," Casgar said.

On one occasion, Huizenga and Casgar were negotiating with an owner in New Orleans whose office was above a malodorous waste transfer station, not uncommon in the business. The facility was equipped with an air freshener system that would periodically spray the sweet scent of a flower. "You can imagine the smell," said Casgar. "We'd sit there and almost die, then you'd have a flower-like smell pervading the office. You'd spend the whole day there." Their patience—and olfactory tolerance—paid off: They ended up signing a deal.

Waste Management used a standard contract to complete the transactions. The mergers would mean the acquired company's lawyers were going to lose significant clients. Dealing with various lawyers—with styles different in the North and the South—took some tact. "We needed to have certain terms to roll in under the SEC rules," Casgar said. "To roll it up and include their numbers in our numbers. The size was important and the kind of volume and work they did. We'd work on the contract and get it signed."

While there may not have been a five-year plan that a cohort of MBAs had developed, the Waste Management teams followed a fairly defined approach to review and reel in new companies. It essentially consisted of three parts: marketing, operations, and finance.

STEP 1: THE MARKETING SWOOP-IN

The marketing team, led by John Melk, analyzed potential markets that the company was interested in entering and competing in. Ideally, the effort focused on a county or group of towns that had a population of 250,000 or more. The more people, the more waste they generated and, of course, the better the opportunity.

Melk had a few people on his team, including Dave Blomberg, who also had been Buntrock's assistant, and Dave Jorgensen. Together they would identify the biggest player in a market as their target. They would swoop into the target's market to look things over. Jorgensen "ran around peeking over fences," Melk said. "The surveillance team would go out to some city, find out who the biggest [company] is and then do initial reports on how many trucks [they had], and this and that. We'd start a file on all those, and we ran through all the data that you could get without computers at that time. So we had America as our market at that point."

STEP 2: GARBAGE OPS

Once the marketing team identified the merger prospects, a second team of operations people would join the effort. The hands-on manager Phil Rooney, then senior vice president, participated in and directed the operations team, which was then comprised of one or two people. They would also go to the location and analyze from an operational perspective.

Don Price, one of the first managers to join the company in 1970, was key to leading this effort with Rooney. He knew his way around waste hauling operations. He led what essentially were reconnaissance operations of prospect companies. He and his cohorts believed their mission was to catch and surpass BFI in annual revenue, which had had a two-year head start on consolidation efforts after going public in 1969. During that time, BFI was concentrating on acquiring larger companies, while Price's group chased solid but smaller businesses, he said.

"BFI was after the big companies, with fifty trucks," Price said. "We were also competing against SCA Services, which was also pursuing smaller operations." To accomplish their mission, Price and team traveled near and far. They were on the road five days a week, hitting towns and sometimes four cities in a day. They believed they were in a horserace with BFI.

They approached their prospects informally. They worked to put the owners at ease. They dressed casually and avoided wearing suits. They'd meet the owners and discuss the business and the future of garbage collecting and disposal. The meetings weren't intimidating, Price said, because Waste Management's operations crew were comfortable talking about their shared experiences. They were good at it—they were operating guys. They could identify with the smaller owners. They'd kick tires and talk trucks, containers, and the everyday business of operating a hauling company—a way of relating, they believed, that BFI could not match, Price said. "It was easy to win people over when you'd go in with a handshake, talk business, tires, trucks, ride routes, and meet the guys. We went in with experience behind us."

"Most of those guys started with one truck and grew the company. They were hands-on kinds of operators," Debes said. The Chicago companies, which tended to be larger, were more sophisticated. But many of the operations the company acquired elsewhere were not. When teams got to visit these operations, it was "casual, casual, casual," he said.

The reconnaissance teams peeked over fences, too. They'd deliver their reviews on Saturday mornings at the company's new headquarters on Jorie Boulevard in Oak Brook to Buntrock, Huizenga, Flynn, Rooney, and Melk, who would then decide the next steps.

STEP 3: THE FINANCIAL REVIEW

Talks with the targets would ensue, and then the third element of review would occur: The financial team would arrive to scour the prospect's books, undertake "purchase investigations," and generate reports that would be the basis for the price offered. Almost always, Waste Management was offering stock for stock.

Debes and company would arrive. "When we were first starting, we—all four of us—were on the road at least three out of five, sometimes five out of five days a week," Debes said. "I'd write my reports on Saturdays and Sundays, and I'd bring them in on Monday. We'd review them and get them to Flynn so he could get them off to the group that was approving them. Depending on the size of the deal, Flynn would send it either to the board of directors or to the executive team. In many cases, it was sent to the officers and they came up with an offer."

The team members were equipped with "write-your-own-airline-ticket" books, which allowed them to go up to a flight counter, snap off a ticket, write down the flight, and get one to the next town. "I'd be in the airport with my carry-on bag, standing in line getting on a flight," Debes said, "and then I'd get a call and they'd say, 'Don't go to Cleveland, we want you to go to Columbus.' So I'd get out of line, cancel that ticket, and go to another desk and get a flight to Columbus." His suitcase was always nearby, always packed. He had to be ready.

The financial review teams seldom stayed in a city for more than a day and a half. They'd rely on accountants for certain records. "Most of the companies had only internal bookkeepers, but they didn't have controllers because the companies were too small," Debes said. A tax preparer or CPA would be pulled in to prepare financial statements. The larger companies had controllers.

"And then I'd go in there and get to know everybody and say how wonderful we are, and we'd like to buy your company," Melk said. "I'm a fuzzy-cheeked kid running around closing deals with older guys who built the business."

EMPOWERING REGIONAL COMPANY OWNERS

The concept worked, Melk said, because Waste Management brought capital that freed the owners from having to find financing to buy new equipment in order for them to grow. "And Dean had a good reputation as well as Wayne in the industry," Melk said. "It was the concept. It was an industry waiting for consolidation and financing. Going public gave us the ability to do all that. We'd identify a market we wanted to be in. We'd pick out the best markets to go in and find the best companies. Then we would say, 'You know us from all the industry meetings and you know Wayne and Dean and how successful they are. We're now a publicly held company so that we can raise money, and the stock is going up.

"'We have all the management skills, the purchasing capabilities with the truck manufacturers and the body manufacturers, like Heil or Leach or Dempster and all those things, and so instead of you figuring out whether you can put another container unit on for two hundred grand, how about [adding] six? We can help you grow, and we're going to do that. We'd like you to be on board and run it in this market.' It gave them an opportunity to grow. That was really our big pitch. You could really grow. Our stock value will grow faster than the value of your company. We want you to be on board, and we're the industry leader. You know us."

Waste Management was buying assets, yes, but they were also assembling an experienced operations team that added talent well beyond the corporate headquarters. The business was on the streets.

And Waste Management nearly always wanted the owners to stay and expand their newly entered market positions. The acquired owners knew their businesses, their capabilities, and where the opportunities lay. They were top operators whose know-how and networks would help the young company grow.

Melk estimated that 80 percent of the owners remained and became division or regional managers. They knew the issues their businesses faced and how difficult it was to get financing and to grow. They liked what they heard from Waste Management.

THE HUNT'S LEADER

Huizenga became the team leader in the acquisitions hunt. He had started his own hauling company with a single truck and leveraged this experience. He was "a fantastic salesman," Debes said. "'Your garbage is our bread and butter,'" Huizenga would tell prospects, Debes recalled. "Once he had his blue eyes locked on you, he had you right in the palm of his hand."

While Huizenga's approach to growth was aggressive, he could temper it. He knew what to look for and how hard to work for it. He could be both focused and philosophical at the same time. "He told me one thing," Debes said. "'Always know what you say no to. Don't pass up an opportunity just because you don't think it's worthwhile. Research it to make sure that you're right about saying no to the deal. Always know what you're going to say no to.'"

"I think we bought over one hundred companies with Wayne and me on airplanes," Melk said. "We'd start out and go to Cincinnati, have a meeting and go to Boston and have another one, end up in Toronto and then have to go back to Cincinnati because BFI came in and upset the deal. It was constant travel and hard work.

"On Saturdays, we'd meet in the conference room and price these companies out and make offers. Then we'd have to hang onto them sometimes for a month or two waiting for the shelf registration to become effective. Then we had to run around and close them all at once. In the meantime, we had to keep BFI from sliding in there and upsetting the deal. It was pretty competitive, very competitive."

THE STOCK PRICE RISE

As the mergers added up, so, too, did the earnings. The accounting treatment could take several forms but typically purchase and merger accounting was used. Both provided control of the target company, but one method, called pooling-of-interests accounting, offered superior results in reporting balance sheet values and immediate earnings results.

The pooling-of-interests method was the recognized standard then applicable to and appropriate for mergers. Waste Management was exchanging its shelf-registered shares to acquire the stocks of the target companies. The approach was favored by the target company owners because, for tax purposes, they would not be taxed on their gains until they sold their Waste Management stock. Each target company exchanged its stock for Waste Management stock. With pooling, the target's balance sheet was added to the parent company's, and historic profits and losses were added to Waste Management's results as if they'd always been part of the company. This generated an immediate boost to profits. As more earnings were generated, the company's stock value rose.

At one point in the early 1970s, Waste Management's stock enjoyed a forty multiple, exciting investors and the former owners who had merged their companies and bought shares. They were delighted to be growing wealthier as they saw the value of their shares ascend. While the stock price rose, they knew that their merger agreements placed restrictions on them if they wanted to cash in their shares. Should they wish to do so—and there were a few who did—the Waste Management generals had gifted them a handsome reminder: They received sets of gold Cross pens mounted on a base with a gold plate etched with the date of their closings. On each anniversary date, the owners could sell only 20 percent of their shelf registered stock.

The added operations began to roll in. In its first twelve months, the company succeeded in bringing in more than one hundred companies. Wherever they turned, however, they were competing against BFI. The executives were always worried, Melk recalled, of "BFI sliding in there" and outbidding them. Melk even wondered if BFI somehow knew how to track their plane travels and the stops they were making. "I'd leave, and they'd come in," he said.

Despite Melk's worries, BFI was not tracking their flights or monitoring their day-to-day whereabouts. They were just fierce competitors who wanted to be at the head of the prospect's line. "We did try hard to get there first," BFI founder Lou Waters said. "It's always easier to acquire by getting there first, explaining [to the prospect] why we were the best, and negotiating a fair price. When we did run into each other, it did cause the price to escalate, sometimes too much when Huizenga was involved. Buntrock was always the more cautious and reasonable one."

TEAM MEMBERS LISTED IN THIS CHAPTER

Ken Arnold	Dennis Grimm
Larry Beck	Peter Huizenga
Dean Buntrock	Wayne Huizenga
Amy Burbott	Al Morrow
David C. Coleman	Bob Paul
Don Flynn	Bill Schubert
Bert Fowler	Peter Vardy
Hal Gershowitz	Don Wallgren
Jerry Girsch	Jane Witheridge
Joe Graziano	Greg Woelfel

9

ENVIRONMENTAL INNOVATORS

The company's early years were formative in building its success and foundation. And one of its most instrumental contributors was Hal Gershowitz, the company's senior vice president, who oversaw corporate public affairs and government affairs, and served as the senior spokesperson until the early 1990s.

Gershowitz was the public voice of the company. He told the company's story, presenting to Wall Street and shareholders and acting as the media contact and legislative liaison, among other external responsibilities. His role couldn't have been more important. And he was perfect for the job.

Before arriving at the new company headquarters in Oak Brook, he had spent six years—most of them working closely with Buntrock—as executive director of the National Solid Wastes Management Association.

Gershowitz brought energy and leadership to the association and, among other things, directed its 1972 comprehensive study with the EPA into America's increasing solid waste generation and the private sector's critical role in managing it. Always looking ahead, the industry's leaders had needed a stronger voice in Washington, DC, to keep tabs on any forthcoming regulations and, if they were in the legislative inbox, to help shape them. It was why they had hired Gershowitz in the mid-1960s.

Waste Management's leaders had been impressed by Gershowitz's management talents and understanding of public policy. And they wanted him to manage all the company's "outside" requirements.

Recruiting Gershowitz had been a competition. BFI wanted him, too, and they were willing to pay quite handsomely for his services. The Waste Management team in Chicago needed to persuade Gershowitz that they would treat him well. So Buntrock and Huizenga traveled to Washington to persuade him. They said they'd make it worth his while. They did. Waste Management was relieved to have him.

On October 1, 1972, a year after Waste Management's public offering, Gershowitz joined as president—Huizenga's idea, Buntrock said. Gershowitz assumed many of Buntrock's administrative duties, and, as the company told stockholders, he would "represent Waste Management in our increasing involvement with the public sector of the solid waste industry. We are now free to devote increasing portions of our own time to operations and planning."

In its earliest days, Waste Management was willing to take on big opportunities and challenges. Gershowitz was sometimes involved in internal developments, too, but they were exceptions to his critical external role of addressing unending issues with governmental bodies, which often had potentially huge projects associated with these relationships.

INNOVATIONS IN WASTE MANAGEMENT

Gershowitz was, for example, a leader in developing a first-of-its-kind recycling plant in New Orleans called Recovery I. The plant was designed as an environmentally safe way to recover metals, paper, and other scrap materials for resale to industry. Residue from the process would then flow to a nearby Waste Management disposal facility.

Recovery I was "dropped in my lap the first day I came to the company," Gershowitz recalled. The company had a few waste hauling contracts in the New Orleans area and learned that BFI was bidding on a resource recovery, or recycling, project that all of the city's waste would go through. The project was being coordinated by the nonprofit National Center for Resource Recovery, which included some high-profile businesspeople and NASA officials.

"I was asked to head up a program to see if there was any way we could get the contract," Gershowitz recalled. "It took a good part of a year to negotiate with us going down to New Orleans to negotiate." BFI was doing the same thing, he said. "We were going in and out like a revolving door."

During his final presentation, Gershowitz sat next to astronaut and first-man-on-the-moon Neil Armstrong, who had endorsed the company's proposal. Soon after, Waste Management won the contract and built the recycling facility near its landfill. The property was adjacent to a bayou and regularly visited by alligators.

Waste Management, the City of New Orleans, and the National Center for Resource Recovery shared profits from the innovative program. The company never felt it could make money from running Recovery I, however, "We used to joke about what we wanted to name it," Gershowitz said. "Don Flynn suggested Chapter 11."

In the early 1980s, Gershowitz was again at the center of marketing a major project—building the company's first waste-to-energy plant in Tampa's McKay Bay. The company had established a relationship with a Danish company to use its System Volund waste-to-energy technology. The experienced Danes burned residential garbage to generate steam that was then delivered through grids and sold back to them to provide heating. Gershowitz and financial executive Jerry Girsch traveled to Denmark to negotiate for the technology, and after winning the contract, created a joint venture company to build the plant in Florida. An experienced builder, Ted Szoberg, was brought on to manage construction of the plant, which was the largest in the nation at the time and sold electricity to the Tampa power utility. The project would not have been possible without Waste Management's relationship with the Danes.

WASHINGTON AND WALL STREET PROWESS

In addition to his talents as an executive and promoter of company innovations, Gershowitz understood lobbying and organizing to support or oppose rules. A University of Maryland graduate, Gershowitz had grown up in the nation's capital and knew how Washington worked. Early in his career, he had managed an underdog's successful congressional campaign in Maryland. He knew how politicians thought, and he ultimately would become the architect of Waste Management's government affairs capability.

Gershowitz was also an expert in public relations and dealing with the media, handing out meaty answers that reflected well on the company. He was a valuable advisor to the operators in the field in handling the frequent sensitive issues that would pop up. He was great at telling the company's story, particularly when talking to reporters. Among his many gifts, though, was his calming eloquence. With storms constantly swirling, Gershowitz was in the public eye thoughtfully, rationally, persistently articulating the company's positions. A better corporate spokesman there never was, his associates would say. Oratory was his forte. Company executives later shared that they hated to follow him in making presentations, lest their performances be compared.

But serving as spokesperson was only part of his role. Gershowitz also helped present the company's story to Wall Street, a role that had not been planned but unfolded naturally. After a few years, Buntrock reassumed the role of president, and Gershowitz was named a senior vice president, responsible for developing resource recovery projects and corporate and government affairs, in addition to being the senior spokesperson. He would have this title until his retirement in the early 1990s. In that role, he and Flynn were the company's ambassadors to the investment community.

The two began meeting with financial analysts. This was new to Gershowitz. Presentations, however, were not. Before long, he and Flynn were flying to Paris, Geneva, and Japan to showcase the company. They were accompanied by the other most senior executives. The teams were assisted by The Chicago Corporation, the investment banking firm that had taken them public, and later by White Weld & Co., which would merge with Merrill Lynch and in the process establish Merrill's investment bank capability. H.W. "Brick" Meers led the White Weld team supporting the company. Meers's assistant was a young Charles A. "Chuck" Lewis who would eventually lead the company's relationship with Merrill Lynch.

"They gave us a lot of respectability in the financial markets, respectability to the extent that financial markets knew this garbage company was someone they had to know and talk to," Gershowitz remembered. They would do a financial tour once a year and then started traveling to Europe every other year. At first, they'd attract a handful of analysts sitting around a conference table having lunch or dinner. It wasn't long before the analyst numbers increased and larger meeting rooms were required.

A TURNING POINT TOWARD GREEN

As the company busied itself securing new investors, it simultaneously needed to keep pace with the nation's growing environmental regulations—and ensure its landfills were in compliance.

Landfills were highly sought, highly valuable assets for the company from its very beginning. The company's CID facility on Chicago's South Side was one of the impressive assets bolstering its initial public offering. Since then, Waste Management had been on a tear adding more to its inventory. It designed new facilities and updated old sites. There was much work to do.

During the 1970s, amid growing public outcry over the nation's uncontrolled dumps and the contamination they caused, the government began working on new rules to ensure their safety. The Solid Waste Disposal Act of 1965 had given way to the Resource Conservation and Recovery Act (RCRA), enacted on October 21, 1976, and signed into law by President Gerald Ford. It empowered the EPA to impose strict regulations on managing and disposing of solid and hazardous wastes.

For companies with resources, RCRA meant opportunity. For those without, RCRA was a major threat. Waste Management saw the former.

In 1976, during a company retreat on Captiva Island, Florida, Gershowitz made a compelling presentation with display boards to the company's management team about these opportunities. "It showed all of the areas of new legislation that represented opportunity for the company," he recalled. "The money that was being allocated in each area that RCRA was covering, all of it meant enormous opportunity for us."

The meeting represented a turning point for Waste Management. "This was when we first embraced tough environmental standards as an opportunity more than a threat," Gershowitz said. "This was where and when the company's solid commitment took root to evolve a first-class commitment to tough environmental standards."

They recognized the upside. "The company was really very excited about the opportunity compared to the threat of RCRA," Gershowitz said.

Waste Management wanted someone to ensure that their facilities would be the best going forward, that they would meet future environmental safety requirements, and avoid liabilities. Buntrock wanted capability inside the company to be ahead of the regulatory curve. He wanted the company to be the industry leader in promoting best practices.

In early 1973, Waste Management was on the hunt to find someone to tackle this job. The company wanted and needed credibility before regulatory bodies.

THE LANDFILL PRODIGY

Peter Vardy was a pioneer in landfill design. He had earned a national reputation in designing environmentally secure disposal facilities and helped Waste Management's landfill sites be second to none. Vardy's work ushered in highly engineered environmental facilities, not dumps, a word the company deplored and considered offensive. These facilities would be impressive showplaces the company could advertise to develop new ones.

Vardy was born in Romania and raised in Europe and Israel. He moved to the United States and earned his degree in 1955 in geological engineering at the Mackay School of Earth Sciences and Engineering of the University of Nevada, Reno, where he had a scholarship. Vardy had a brilliant mind. While in school, he'd been a poker player in the casinos. He could read the cards and beat the house until the introduction of the multi-deck shoe changed the odds to favor the house. He spoke six or seven languages. He had even been a dance instructor. Vardy was called back to military duty in Israel during its 1967 war, where he served as a demolitions expert on a mine sweeper.

After earning his degree, he went to work at the engineering firms Dames & Moore and then Cooper, Clark & Associates. It was at the latter company that he designed the nation's first fully engineered sanitary landfill, a modern facility where waste is disposed of in an area isolated from the environment, in this case, at a 544-acre Mountain View site in Santa Clara County, California. Vardy then founded EMCON Associates to expand this approach nationally. Within two years, the firm had designed 120 sanitary landfills. Along the way, Vardy met his wife, Lillian, in New York. By coincidence, she was from the same Romanian town he was, Timisoara, and had been a Nevada casino dealer.

Gershowitz was dispatched to recruit Vardy. They met in San Jose, California. Vardy had put his all into building EMCON and wasn't interested in leaving it. But Gershowitz made a proposal: Waste Management would bring EMCON and its engineers into the company. Vardy accepted and became the company's vice president of environmental management and technical services, essentially its chief engineer.

"We worked it out as a way for him to have a hand in both operations. But Peter then left the operation of EMCON to others, good people, and he became our first professional chief engineer," Gershowitz said. "And he gave us a lot of credibility too, because Peter could discuss disposal site engineering as well as anybody in the government."

Vardy was actively adding to the environmental management team. Jane LaPorte Witheridge was one of Vardy's early recruits.

A YOUNG ENGINEER JOINS THE TEAM

Witheridge was twenty-two in 1976, an engineering graduate from Lafayette College in Easton, Pennsylvania. She knew a little about solid waste after having interviewed an author of a book about refuse collection productivity and writing her senior thesis on waste. And she was looking for a job.

Vardy brought her in for an interview. She declined his offer at first and went to work for an environmental consulting firm in California. But she soon discovered that she couldn't possibly consult without really knowing the solid waste business. She called Vardy back and asked if he had filled the position. He had but said there was room for one more engineer. Witheridge returned for an interview.

"The first person I met with was Larry Beck on Jorie Boulevard in the boardroom," Witheridge recalled. "And it was like *Mad Men*. There was a little tea trolley out in the hallway. Larry was sitting in the corner dressed casually in jeans and wearing his snakeskin boots. That was my introduction to Waste Management." Beck liked her and quickly passed her along to Vardy.

Witheridge was familiar with the company's Recovery I New Orleans resource recovery project. She had read about it the night before. She next found herself with Vardy at the big CID landfill on Chicago's South Side being introduced to the maintenance shop workers. She was dressed for an interview in a nice skirt, jacket, and high-heeled shoes. Not exactly the attire for a day at the landfill.

"He [Vardy] took me out to the tipping [disposal] area," she remembered. "I had to climb up a berm in my high-heeled shoes. He offered me his hand, which I refused to take. I thought, *If I'm going to be with you, I'm going to be able to climb up this stupid berm by myself,* which I did. And when we got to the top there was this mass of garbage like fifty feet down and seagulls and trucks and all kinds of stuff. And he said, 'Take a deep breath.' And I did, and he said, 'What does that smell like?' I didn't know what to say. He said, 'I'll tell you what it smells like. It smells like money.'"

Witheridge got the job, her introduction to a nearly twenty-year career with the company.

It was a position in a then-man's world, and some senior managers later said they thought she'd be challenged dealing with the men of garbage. She more than held her own.

EXPANDING ENVIRONMENTAL OVERSIGHT

Two weeks after Witheridge joined the company, the RCRA legislation passed, marking the start of national standards for landfills. The federal RCRA law meant more capital was invested in landfill facilities, as synthetic liners, leachate collection systems, methane recovery, and other environmental controls and financial guarantees were required. Over time, many municipal landfills closed their gates. In 1970, the nation had about 20,000 landfills. Two decades later, there were fewer than 3,000. Most would not meet the standards. "The Resource Conservation and Recovery Act forced the improvement of standards and it required large companies then to invest in those standards. It was a perfect confluence of being able to have a strategy and related acquisitions with a regulation tailwind behind you," Witheridge recalled. There were only a few other technical engineers working on landfills then, Greg Woelfel in Michigan and Wisconsin, Bert Fowler in Illinois, and Jerry Gresh in Florida. There were almost no women. Witheridge traveled constantly and, like so many others, was armed with write-your-own-plane-ticket books. Waste Management was keeping the airlines busy.

Witheridge would cover a lot of ground in the years ahead, earning different assignments and responsibilities as issues arose and the company matured. She helped to pioneer the company's environmental audit program, modeling it after the financial group's internal audits, to ensure that things would be done right at the operating sites. She worked internationally on projects in Saudi Arabia and South America. She reviewed site permits to make certain the sites were in compliance. She ran regional landfills and later led the company's recycling program.

Initially, though, Witheridge conducted reviews on landfill acquisitions, standardizing inspection protocols. Over time, she would review landfill acquisitions totaling more than a billion dollars in value. She found herself frequently accompanying Vardy to the boardroom, where decisions were made. She did much of the talking. "I knew the regs. I knew the site. I knew the permit," she said. She knew the details. "Every little non-science issue was a matter of record."

When RCRA rules began to come into effect in 1980, Witheridge directed coordination of the site permit applications. She reported on it to Bob Paul, updating him weekly with three aides typing up the updates. It was a ton of work. "I carried all the applications to Peter Huizenga, who was secretary, for his signature. I actually got a bonus for my work. It was really gratifying."

Witheridge said she liked the idea of working in the environment and the dignity it offered to people doing an honest day's work. "It resonated with me pretty deeply," she said.

In 1978, Witheridge became the regional engineer for operations in the South and Southeast. She reported to Al Morrow, who oversaw eastern, southern, and central states. He was a stickler, a master of detail. "He would count the number of telephone rings you had before you picked up the phone. I loved working for him because he was very clear about what he wanted. If you weren't going to meet a deadline, all you had to do was tell him why and tell him ahead of time so he had no surprises," Witheridge said. There were new greenfield sites to develop and expansions of existing facilities. There were some sites with problems, most with promise.

In her new role, Witheridge provided support for local zoning meetings and presented. At one Texas hearing, a public official mistook her male colleague as the engineer. "It was the first time he'd met a woman engineer," she remembered. She was then promoted to the district vice president, responsible for landfills and transfer stations in Pennsylvania and New Jersey, reporting to Region Vice President Dennis Grimm, a get-things-done kind of manager.

Witheridge and a group of environmental engineers helped sites in Pennsylvania expand—an area of value for the company (more of which is described in greater detail later). At the time, there was a regional disposal crisis.

She worked closely with Grimm and Joe Graziano, an expert in responding to requests for proposals, and lawyers Amy Burbott and Ken Arnold. They helped open Tullytown, a new landfill, and transfer stations known as Avenue A in Newark and Essex County. The team also added its Forge and Philadelphia Transfer facilities, sites where residential and commercial collection trucks head when fully loaded to drop off their tons of compacted waste. Transfer stations, garbage's whistlestops, are busy, fast-moving operations. Safety is critical. At these sites, operators of big-iron, rubber-wheeled loaders push the material into piles to be scooped up by mechanical grappling devices and deftly placed in large semi-trailers for shipment to distant landfills.

Rooney and the executive team had placed high value on transfer stations and landfills and filling out their disposal facilities network. They were needed because disposal sites were becoming farther removed from urban centers. As landfills dispersed, it no longer made sense to take collection trucks off their routes for long trips to distant disposal sites. They were needed on the streets. Productivity mattered.

After Witheridge left Pennsylvania, she went back to the corporate office. It was time to make sense of recycling, a critical building block of the company's success.

PRIORITIZING COMPLIANT LANDFILLS

Vardy's environmental engineering group continued to grow as engineering and developing landfills that met the regulations were critical for Waste Management. In 1983, Vardy recruited Don Wallgren, a senior environmental engineer from the EPA's Midwest region, to be the company's vice president of environmental management, focused on the company's North American disposal sites. Wallgren was well-known across the EPA and the industry and had written a number of technical papers on landfill design. His job was to develop a team at the corporate office and in the regions to ensure the company's solid waste disposal facilities were up to par and in compliance with rules current and in the future.

"Don was brought in by Peter Vardy to develop an engineering staff," said Bill Schubert, the company's longtime vice president of engineering in the Midwest. "He probably looked like a perfect fit because he was an engineer with a US EPA background. However, Don's real value in the corporation was his ability to communicate with people at all levels." Wallgren was a Minnesota boy with a natural Midwestern humility in his business deals and was respected by everyone, Schubert said.

Jerry Girsch agreed. "Don would deal with the issues in a very pragmatic fashion, and we'd get results and move on," he said. "Don brought a lot to the organization. We had people reporting to Don in every region. Every region had a regional engineer. As the organization grew, these functions couldn't be handled out of corporate, they needed to be handled in the field."

TEAM MEMBERS LISTED IN THIS CHAPTER

Doug Allman	Ron Jericho
Dick Ancelet	Bill Katzman
Ron Baker	Jim Koenig
Larry Beck	Jerry Kruszka
Tom Blackman	Jeff Lawrence
Bob Brach	Richard Leitzen
Dean Buntrock	Don McLaughlin
Don Clark	John Melk
Mike Cole	Dick Molenhouse
Tom Collins	Al Morrow
Jack Cull	Jay Nowack
Bill Debes	Oscar O'Bryant
Jim DeBoer	Dave Pearre
Earl Eberlin	Don Price
Rich Evenhouse	Paul Pyrcik
Ed Fixari	Fred Roberts
Don Flynn	Mike Rogan
Jim Gencauski	Jay Rooney
Herb Getz	Phil Rooney
Jerry Girsch	Stan Ruminski
Dennis Grimm	Jack Sher
Fred Hahn	John Slocum
Bob Hanson	Harold Smith
Tom Hau	Leonard Stefanelli
Whitt Hudson	Eddie Van Wheeldon
Wayne Huizenga	Peter Vardy
Bill Hulligan	Jerry Veach
Joe Jack	Don Wallgren

10

THE UNCEASING HUNT FOR
ACQUISITIONS

The first years after the public offering saw Waste Management consolidate hundreds of companies. Revenues and earnings were climbing rapidly. In the span of three years, revenues had grown from $16.8 million to $158.5 million by the end of 1974. That year, Waste Management had nearly one hundred operating divisions, almost one million residential customers, and 117,000 commercial customers.

The rollup was well underway, and Don Price was a very busy man. He and others were part of operations teams dispatched by Phil Rooney across the country. Their mission was to size up hauling and disposal companies for potential mergers. They also made sure any recently acquired operations were performing as expected.

The acquisitions teams were well-coordinated. Price worked closely with Rooney. They were armed with a list of contacts of smaller companies from the National Solid Wastes Management Association, the Detachable Container Association, or other friendly relationships. Price and his team would use these leads to travel to new markets and look the prospects over. They were looking for residential, commercial, and industrial routes.

"Once I was at a location, I'd actually use the *Yellow Pages* to find other companies to check out," Price recalled.

Other operations team members, such as Ed Fixari, were climbing into planes, commercial and private, and rental cars to reach the locations where they could get a close-up look at the operations Waste Management was considering. They flew and drove at night. Daylight was for doing business.

Debes had responsibility for operational and financial reviews and focused mainly on states south of the Mason-Dixon Line except for Florida, which was Huizenga's special purview. "In the South, it took two or three meetings before they'd accept. They needed to check us out. It was really a slow process," Debes said. "They had to establish a relationship before they opened up. If Wayne gave anyone a verbal offer, that was what they got." Also on the acquisition team, assigned by Huizenga, were Jack Sher and Oscar O'Bryant. Sher was based in Spartanburg, South Carolina, and pursued leads from Virginia to Alabama. It helped that he had his own plane. O'Bryant had Louisiana and the West, excluding Arizona and California. "Whenever I was assigned by Wayne to talk to a prospective acquisition candidate, Jack and Oscar had them warmed up. Both were great goodwill ambassadors for Waste and our acquisition efforts," Debes said.

THE NONSTOP HUNT FOR COMPANIES

The hunt never ceased. The marketing scouts were organized. They had to be—there was so much to keep track of. Price kept a neat calendar with square-inch blocks that he would fill exactingly with his schedule, the cities he needed to visit, the prospect company's name or owner, the things they needed to address, and the problems they needed to solve. He'd squeeze in all the details. It was a daily diary of his business life. He even kept travel notes on clothing, food, and motels. It was "very, very busy," he recalled. One week, he would leave on a Sunday night to go to Florida, beginning in Orlando, then Jupiter, then back to Orlando, and then home late Friday. The next week he'd make stops in Ohio, and the following week, Washington, DC.

In the first week of January 1973, for example, Price's calendar had him in Wilmington, Delaware; Washington, DC; Columbus, Ohio; Hartford, Connecticut; Philadelphia, Pennsylvania; and back to Washington. The following week he was in New Orleans, Louisiana; Spartanburg, South Carolina; Dallas, Texas; Erie, Pennsylvania; Topeka, Kansas; and Springfield, Massachusetts. These days on the road were common to all of those searching for companies to talk to.

"Wayne Huizenga was driven to do as many deals as he could and set the pace," Price said. "We read routes to get a flavor for the market, the shape of the equipment, and the attitude of the drivers. You would be surprised at all the information you could gather from a driver by lunch break."

One of the first companies the team attracted was Benton Disposal in Toledo, Ohio. Benton's assets were split into two companies, one owning the garbage trucks and the other the waste containers. This was unusual, but the teams figured out its worth and how to get the merger done. It would be the first of hundreds of acquisitions over the next few years.

Sometimes the decision-making could go quickly. Some deals were done seemingly overnight. Some complicated the valuation process. In one case, Price and Debes met with Don Clark, an independent operator who had a three- or four-truck operation in Jefferson Parish near New Orleans. The company was eager to do the deal, and Clark was entirely receptive to its overtures. After a few phone calls with Flynn, Huizenga, and Buntrock, and only a few days of back-and-forth, they had an agreement. The papers were sent, the deal inked.

Recruiting Clark and his company was a valuable move. Clark was a dynamic salesman. Blessed with his ability and armed with Waste Management's resources, he would grow the operation to become the largest hauler in New Orleans. The acquisition served as a model in how Waste Management could expand its local operations once it gained market entry.

There was a method to their ways. Throughout this time, Price learned more about acquisitions and the importance of being decisive from Rooney. He and Rooney hit it off well. Rooney was a leader, Price said, and for him, decision-making came easy and without fear. Rooney was also on the road five days a week and related well to the owners he and Price visited. "Every week it grew, we had a new company or two," Price said.

Not all of their efforts were focused on mergers. They also had situations to repair. Rooney, Price, and Jerry Kruszka, who, like Price, had started working with Stan Ruminski at Acme Disposal in Wisconsin, were working to solve landfill problems. At one problematic disposal site, the three of them sat in a Columbus, Ohio, hotel room trying to figure out the best way to clean up the operations.

"It was a mess, and the landfill had old equipment," Price remembered. Rooney didn't hesitate. He picked up the phone and placed an order for a new Caterpillar bulldozer and compactor, not inexpensive items, to be delivered to the site. "For a young guy, Phil had a presence and a command about him." Kruszka ended up staying in Columbus for seven months, becoming the site's general manager. Afterward, Rooney assigned him to go on to repair and manage other problem sites. Kruszka would enjoy a long career as a senior operations executive in both the company's solid and hazardous waste divisions. After leaving Waste Management, Kruszka would lead and grow other important waste companies.

At times, there were odd instances in work to win Waste Management over. Once in Louisiana, a Waste Management team was taken out to a rural area and treated to Southern barbecue, coleslaw, and deep-fried catfish. Afterward, the two prospects' offer to introduce them to female entertainment was politely declined. They were turned off, but the deal would eventually get done. However, those owners would not be joining the company.

WHEN MERGERS WENT ACCORDING TO PLAN
In Chicago, Dick Molenhouse, who owned Molenhouse's Clearing Disposal in the suburb of Oak Lawn, was among the first to join Waste Management, rejecting other suitors. His company did commercial hauling downtown. Molenhouse had grown up with Huizenga. They

were "friends forever," he said. He believes he was even with Huizenga when Wayne bought his first truck.

For Molenhouse, and, over the years, hundreds of others, the benefits of merging with Waste Management were clear. Molenhouse knew the value of his business and could figure out its future worth. "I guess we can all count, right?" he would later say. "Would you like to be a part of a hundred-thousand-dollar operation or have 10 percent of $20 million? When you put the numbers together, it made sense." But the decision was based on his business savvy, too, and the risks he understood as inherent in the waste industry. He saw the value of such a merger in providing financial security. "You could unload your liabilities. Every time you sent a truck out on the street, it's a liability." Molenhouse thought he'd stay on as a manager for a few years. He retired twenty years later.

Other companies that quickly joined in the Chicago area included the DeBoer family's business, Garden City Disposal, which had a transfer station and served O'Hare International Airport; Southwest Towns Disposal, one of Tom Tibstra and Larry Beck's companies; and Meyer Brothers in Bridgeview. Many more were brought in from all over the country.

The teams looked west, and in September 1972, completed their largest merger to date with Universal By-Products, a $6-million revenue business that gave the company a West Coast presence. Universal served customers in Los Angeles, San Diego, Lancaster, California, and Phoenix, Arizona. It also added a paper recovery and brokerage business.

WHEN MERGERS DIDN'T GO ACCORDING TO PLAN

Not all of the hundreds of transactions went as hoped, though.

One prospect in Boston was nearly ready to close when the company's due diligence team discovered a barge on the books. Fearing environmental concerns, they decided not to pursue it. Another that stands out was in Kansas City. Deffenbaugh Industries, which had more than seventy trucks and a large landfill, was a big catch. But problems began to surface almost immediately after the closing, and Waste Management did not hold onto the company for long.

"The day after we bought it, he [the owner] wanted it back," Price said, "But he couldn't get it back."

Before long, word reached the Waste Management team that the former owner, Ronnie Deffenbaugh, had launched another company in an alleged violation of his non-compete agreement. They also learned that Deffenbaugh had hidden waste containers in a grove on a twenty-five-acre plot of land. Price rented a plane to "do recon," he said, and discovered four containers hidden among the trees during the flyover. Price wanted to get a closer look. His stealth reconnaissance unit, including John Slocum and Bob Hanson, crept up to the site. Trying to conceal their mission, Slocum and Hanson wore black hats and shirts and brought crowbars to pry open the doors of the containers.

"We waited until it was pitch-black," Price recalled. "We almost got to our target when we heard dogs barking and the barking got louder with each passing second. We made a quick retreat, but the barking got louder. Fortunately, we outran the dogs to the edge of the field and the dogs stopped. That venture is still in our memories."

Deffenbaugh soon quit, taking with him drivers, mechanics, and helpers. He also sued Waste Management when the value of his stock holdings declined during the 1973 recession. The company countersued, alleging he violated his non-compete requirement. Eventually, it would all be settled.

Price moved his family to Kansas City and spent the next eighteen months there trying to salvage the business. The Kansas City division became a training ground for up-and-coming managers to search for and manage acquisitions. Among them were Hanson, who came from Acme Disposal in Milwaukee; Ron Baker from Ace Scavenger in Chicago; the detailed Al Morrow, a disciplined new hire with DOT trucking experience; Phil Rooney's brother, Jay; Eddie Van Wheeldon, whose family's Chicago business was acquired; Don McLaughlin from Southern Sanitation in Florida; and Jim Gencauski, who would later work with the international group. They were placed in rented apartments outfitted with cots and helped rescue the purchase. The Kansas City business would eventually be divested only to return a number of years later. For Waste Management, a presence in Kansas City had value. It was a good market.

BUSINESS IN TUSCALOOSA

The rollup was still going strong in 1979 when Don Price met Fred Hahn, a successful trucking and warehouse owner in Tuscaloosa, Alabama. Hahn had gotten into garbage only after a city hospital administrator, upset with his current service, asked him to buy a rolloff truck and install a garbage compactor. The official said he'd pay Hahn enough for the service to pay off the truck and help him get other customers. It was the beginning of a big business.

Hahn's timing was good. He soon had operations in Tupelo and Columbus, Mississippi; in Florence, Jasper, and Sheffield, Alabama; and in St. Mary's, Georgia. His was a growing concern, one coveted by searching consolidators. "The garbage business was exploding," he recalled.

Price alerted Oak Brook. Soon, Rooney and Debes were traveling to Tuscaloosa to inspect Hahn's biggest operation and try to buy the company. Hahn had more than enough to handle with his other business interests and cheerfully entertained their visits. They were "nursing" him, he said. So was BFI. BFI thought it had the inside track to win his business, Hahn felt, until its representative, irate over Hahn arriving late for a meeting, insulted him. Hahn wouldn't have it. BFI would be disappointed.

Hahn liked the Waste Management people, and they liked him. He had one request before he'd do the deal, though. He was concerned about his father-in-law, who worked at the Tupelo business and whose legs had been amputated. "Phil, I'd like to do a deal with you," Hahn said, "but I want to keep Tupelo because of Pops." Rooney grinned, Hahn remembers. "'We'll take care of Pops,'" he said. Rooney knew a good deal when he saw one.

Huizenga had been working the deal, too. He flew into Tuscaloosa and went to Hahn's office. "To show you how naïve I was and everything,

Wayne was sitting in my office over in that chair across from the desk," Hahn said. "He told me, 'Fred, we've all taken a liking to you, and we really would like to see you take all stock.' We'd already decided on a price by then. I said to myself, *Uh, oh, I bet I'm smarter than he is.* So I looked at Wayne and said, 'No, I'll take half stock, half cash.' I don't want to know what that cost me," Hahn laughs at the memory. The stock soared after the deal was closed.

Mike Rogan, then a twenty-five-year-old financial analyst, was investigating an Atlanta company and marveled at the smoothness of that acquisition. "I was with Don Price," he remembers of the time. "He said, 'What do you think?' I said I'm still checking, and everything looks in order. I think we just bought a company."

Hahn enjoyed the Waste Management visits. He would take its executives to Tuscaloosa's private North River Yacht Club on the banks of Lake Tuscaloosa, which had piers and boats and a beautiful golf club. Some in the Waste Management contingent were surprised the Southern city had a yacht club. Once, Huizenga boarded a company jet with other senior executives and announced they were going to stop for dinner at Hahn's club en route to coastal Fort Lauderdale. "'Wait a minute, Wayne,'" Hahn recalls Jerry Girsch, dripping in sarcasm, saying. "'We're in a plane and we're flying to Fort Lauderdale and you're telling me we're going to stop in Tuscaloosa, Alabama, and have dinner at a yacht club.' They liked it so much I couldn't get rid of them," Hahn said. He would stay with Waste Management another fifteen or so years, helping to buy other companies in the South and playing an influential role in the company's Alabama governmental relations.

There were lots of celebrations marking the closings of mergers, often in Oak Brook. One executive remembered chief acquirer John Melk's role as "master of keeping the prospective new manager-owners comfortable" through these friendly rites of passage. The events served as opportunities to meet the company's executives they may not yet have encountered as well as other key people in procurement, insurance, advertising, and finance.

THE HUIZENGA APPROACH

Some members of the senior team were initially concerned about overpaying for hauling companies and made conservative offers. This attitude made negotiations more difficult and meant some prospects got away. After efforts to lure a company or two were lost, they changed their approach. They used company stock and no longer felt constrained about putting it to work. The teams were determined to be aggressive. And aggressiveness fit Huizenga's personality.

Huizenga's competitors recognized this trait in him. BFI's Lou Waters saw it firsthand. Huizenga, he knew, prized pulling in every prospect.

BFI had ramped up its consolidation efforts after buying its first business in Bridge City, Texas, in 1969, one of only two that year. It added a dozen more the next year, and fifty more in 1970.

In 1978, BFI negotiated an agreement to acquire its first hazardous waste business, a company in East Texas. The company was set to pay the concern's five partners $1 million. But Huizenga had other plans. BFI had left the contract for the partners to sign and returned unconcerned to their Houston headquarters. A couple of days passed without word. "So, we called them to see about the contract," Waters recalled. There had been a development. "'After you left that day, Huizenga flew in and offered us a million dollars more,'" Waters said they told him. "'You're paying us $1 million and he's paying us $2 million. We couldn't turn that down.'"

Waters reminded them that they had an agreement. "They said, 'Well, we were concerned about that, we were afraid that you might sue us. But Wayne said he'd fix that.' And he gave them an indemnification agreement and said that if BFI sues you, he'd defend them and take responsibility for the risk." BFI asked to see the documents.

"They showed us that indemnification," Waters recalled. "And there it was, Wayne Huizenga. And then they showed us the [acquisition] documents. I said, 'How in the world could you close that quickly?'" The BFI deal had taken several months. "They said, 'Oh, it was easy. Huizenga just crossed out BFI and wrote in Waste Management.' Now that was aggressive. We went back to Houston and filed suit immediately." The parties would eventually settle. Waste Management would keep the business but pay $1 million to BFI to drop its lawsuit.

A young twenty-three-year-old summer clerk at the Bell, Boyd & Lloyd Chicago office did the research that gave Huizenga the legal support to mark up and sign the BFI contract. That clerk, Herb Getz, would become the company's general counsel fourteen years later.

THE ACQUISITIONS LEADER

One of the company's key players in the acquisitions hunt was Bill Hulligan, who would become one of the company's most senior executives. Hulligan was born into the waste industry. His grandfather, father, and uncle operated three hauling companies in Cleveland. They included Cleveland Maintenance, Metro Disposal, and West Side Disposal, which altogether had nearly forty routes. They also ran a few landfills.

When word came that they were ready to sell, the Waste Management team, including Rooney and Price, approached them and spent several days looking over their business and the benefits a merger might bring. In the end, they couldn't structure the kind of deal that the Hulligan family sought. Instead, SCA Services paid cash for the operations in 1972. Waste Management might not have gotten the Hulligan business, but they'd eventually get one of the family members.

Before the sale, Hulligan, fresh out of the Navy, was hard at work at his family's newly opened landfill while finishing his college education at John Carroll University. He'd go to class in the morning, work at the landfill in the afternoon, and then return to evening classes. He graduated in just three years and then went to work for SCA. He became a manager for SCA in their Michigan and Illinois offices during a turbulent period in the company's history, during which SCA had a revolving door of company presidents. He no longer saw a future at the company and left in 1978.

The day after he left, at 6 a.m., Rooney called to invite Hulligan to work for Waste Management. Rooney was eager to have him on his team. But first, at Rooney's direction, he met with Beck, who would hire him to work in the company's hauling division in Detroit. Waste Management planned to split this Detroit division in half, and Hulligan was assigned to the city's east side while another manager, Jay Nowack, took the west side. He worked under Rich Evenhouse, then the master of Michigan operations. Hulligan performed well. He wasn't there long. A week later, Beck called him. "I need you in Toledo, you are going to be the manager," Hulligan recalls hearing. "'You can move, or you can commute there if you want.'"

The competition for hauling companies in Toledo was stiff, and the company's sales and operations management there needed to change to keep up. In assessing the site, Hulligan found that half of the operation's disposal was going to a city landfill, even though the company had its own site. "I said, 'Why?'" The answer was that it was closer but only by five miles. "When you see things like that, it is easy to make improvements," he said. "Toledo ended up doing well." And so would Hulligan.

His success launched an odyssey. He was transferred to Dallas to be the district manager for three years under Morrow, who was also growing with the company. Next, he moved to Fort Lauderdale to be district manager when Whitt Hudson retired. Then regional manager Earl Eberlin retired too, giving Hulligan another step up the ladder. "Well, I'm leaving," Hulligan recalled Eberlin telling him. "I don't know if there's a whole lot more I can teach you, but I'll be around by the phone if you need me." Two years later, Hulligan was promoted to executive vice president for the North American solid waste business. He was effectively the chief operating officer of Waste Management's North American business reporting to Rooney at the company's Oak Brook headquarters. There was much to deal with.

"It's always that the first issue is people," Hulligan said. "Who do you have, who's doing a real good job, who isn't doing a real good job? If you have people issues, you have growth issues. You have all the issues with running a company: the allocation of capital and acquisitions during that period of the 1980s and into the 90s. Every year we had about eighty to ninety acquisitions. I was involved with most of them one way or another, either final approval, meeting with people over the years, or knowing people over the years." The key to acquisitions, he said, was personality and getting along with people.

The biggest issue Hulligan found, however, was sustaining the growth the company was known for and expected to achieve. Inflation helped. So did its disposal facility network.

"We had infrastructure that was in place with landfills and transfer stations, as well as locations that we had throughout North America, the US, and Canada. And it allowed you to do acquisition tuck-ins nearby. You buy someone who has a better facility, and you had the synergies of a tuck-in," Hulligan said. Acquisitions, new market entry, and tuck-ins, which added routes and customers to existing operations, fueled the growth along with increasing the daily number of customers the hauling operations served.

Several acquisitions, because of their size, were notable. "Oakland Scavenger was probably the biggest one I was involved in," Hulligan said. The business had one hundred and twenty partners. Hulligan and Rooney worked on the acquisition for years, helped with introductions by two company managers in California, Tom Blackman and Leonard Stefanelli. "We had to go to their board, their owners, who were workers as well, and they had to get a unanimous consent for the deal." Each owner received Waste Management stock, which appreciated nicely in the following years. The acquisition was Waste Management's entry into Alameda County across the bay from San Francisco.

Another Northern California acquisition Hulligan worked on was the merger of Empire Waste Management, a pioneer in the early days of recycling, in Santa Rosa, north of San Francisco. He knew the owner, Tom Walters, through the NSWMA. Walters was well-liked and involved in all things business and civic in Santa Rosa.

The company's acquisition efforts worked well, Hulligan believed, because it had capital, a reputation for treating its employees well, good equipment, and fair prices. The owners of the companies they were buying knew those things too, he said. Also, Waste Management didn't use consultants. Waste Management's acquisitions team would review the prospect's financials and envision its future performance under the company. Did they own their facility? Did they have good equipment? Did they have a landfill? Were they in a growth area? Those were some of the key questions. "It didn't take long to do a deal," Hulligan said. "Once somebody wanted to sell and they wanted to sell to you and you wanted to buy it, you could come to a deal."

Hulligan's managerial odyssey would continue throughout his Waste Management career. His titles and locations would change as the organization grew and transitioned and as executives shuffled from place to place seeking further improvement. But no matter his position, Hulligan knew the recipe for the company's success. "It comes down to our employees, the drivers, whether they are pleasant or they're not. The whole success of Waste Management truly is everybody—the guy who fixes the trucks, cleans the trucks, drives the trucks, and works in the office answering the phone."

THE FINANCIAL CONTROLLER

As the company quickly grew, it needed a vast financial infrastructure to match its scale. Jerry Girsch arrived to oversee the expanding network of financial managers.

Girsch was another Arthur Andersen recruit. Waste Management had been his largest client. He, too, was a Flynn acolyte. After nine years in public accounting, the last four years of which he was an audit manager, Girsch was ready for a change. He had mentioned it to his friend Flynn, who was on his way to Saudi Arabia. Don't make any decisions until I return, Flynn told him.

Several weeks later, the two met at the Cypress Restaurant in Hinsdale, Illinois. Flynn was savoring martinis while Girsch sipped gin gimlets. The conversation progressed through dinner and before the check came, Girsch had an offer he couldn't refuse. He accepted on the spot. Returning home, Girsch's wife, Linda, asked him about one detail, "How much is he going to pay you?" "I never asked," he replied. "I just knew Don was the type of guy I would enjoy working for and that he would take good care of me, and he would make it worth my while. That never really changed."

It was 1976, Girsch joined as the company's controller, its chief accounting officer, succeeding Richard Leitzen. He took charge of the robust organization already in place. Leitzen had created a state-of-the-art electronic system to manage the company's accounting and transmit its financial data. Ron Jericho was expertly handling most corporate-level accounting. He didn't need anyone looking over his shoulder, and although he was in Girsch's group, Jericho essentially reported to Flynn.

Girsch focused on the financial results of operations. He spent his time in the field traveling with Rooney, looking at acquisitions, and making sure local managers were attending to the details that delivered the results expected. It could be complicated. There could be a hundred reasons a local unit succeeded or failed in Waste Management's grading system: personnel issues, equipment problems, competition, poor oversight, or a myriad of other causes. Rooney and Girsch attacked the problems ferociously. The battle to be better, more productive, and more profitable was unceasing. Growth. Growth. Growth.

BUILDING THE FINANCIAL STRUCTURE

Although the company's operations architecture would change from time to time, the hierarchy flowed from the corporate office to the regions to the districts to the divisions. Geographies and names might change, but oversight lines did not. And discipline was demanded.

Flynn was the architect of the financial structure, which conformed to Buntrock's plan to have strong, educated financial managers at corporate and in the field. The system had a dual structure: Controllers were installed at divisions to look after the accounting. Division managers directed the operations. The controllers, although working side by side with the division managers, reported up in the organization separately. Flynn's design helped ensure that local managers could not put undue pressure on the accountants.

Flynn also designed a one-page monthly statement that contained the financial results of each division. He could look at the statement and know exactly how a division was performing. "The company was controlled with financial discipline," said Jim Koenig, who succeeded Flynn as the company's chief financial officer and served in that role from 1989 until 1997. "You put a controller in. The controller controlled the cash. There were policies on what could be spent, limits on what could be spent. The controller was basically the eyes and ears of Oak Brook.

"We'd move our guys all around. The controller would work with the former owner, who typically was the general manager, and help him with the bidding and finance and all of that stuff. At the end of the day, cash was moved to Oak Brook. Any big expenditures were approved by the district region. That's how the business was controlled. That was, to my mind, one of the basic underlying principles that made it work. If the general manager was getting out of line, doing something he shouldn't be doing, we had someone there who could see it. That structure really controlled the business. It was ingenious at the time. It was simple."

The company recruited controllers from Arthur Andersen. "They were basically our source of good financial talent to put into the field," said Koenig. Andersen hired talented accountants out of school and trained them, he said. The relationship was of mutual benefit. Waste Management got the financial talent it needed, and Andersen had its people placed in the company, helping to secure the client relationship.

Girsch directed the regional controllers, who supervised district controllers, who guided division controllers, the latter closest to the business on the street where the money was made. He studied budgets and operational efficiency, the latter Rooney's strength. Girsch was a numbers guy, but Rooney had a much keener understanding of operations.

The local controllers managed the general ledgers and supported operating managers. They provided information on being more cost-efficient, pricing customers, advising when to get a new account, and whether money could be made on it.

Budgets were the bible. Controllers found salvation in following the

Flynn financial scriptures and, in Waste Management's case, meeting and exceeding budget numbers. Operating reviews, established in the very early years, were held quarterly and monthly. Budget reviews were at the end of the year. They were known as QORs and MORs. Performance data was collected electronically and processed through Oak Brook's IBM mainframes into formatted reports. The system produced thick three-ring binders packed with all kinds of information. The reports were a financial report card. And grades counted.

"I'd go through the binders and prepare a three-page analysis of what went on in the regions," said Girsch, whose analysis would then be delivered to Huizenga, Buntrock, Rooney, and Beck. Girsch highlighted the results, good and bad.

FINANCIAL REPORT CARDS

Soon after a quarterly or monthly close, the executives had insight on each region, district, and division's performance. "Everybody had to close their books, typically eleven or twelve business days after the end of the month," Girsch said. "All of the information was transmitted to Oak Brook. In Oak Brook, we'd get an income statement of every division. We'd get it the same time the division got it. It wasn't like we didn't know. We knew about it maybe faster than the division preferred."

The report gave current month and year-to-date results versus their budgets. It would roll up from the divisions to the districts and to the regions and then was consolidated at the top. The regional managers and controllers would fly to Oak Brook for the QORs and ready their presentations in the paneled boardroom.

QOR participants would sit around a long conference table made of rich burled wood, its edges separated by a line of inlaid brass. The inquisitors presided, seated before the paneled west wall on leather swivel chairs. They were surrounded by walls displaying rugged scenes of western art. The west wall had paneled doors which, when opened, revealed a small bar, used only for occasions of celebration and cordiality, not budget conferences. The room was comfortable and intimidating at the same time. It was all about business.

The region teams were prepared, they had to be. Many had rehearsed their stories. With the press of a hidden switch, the drapes would close. The projection system would light up. The slides would start. The sessions would begin. They would get down to business.

If there was an issue with a particular district during the presentation, the corporate teams would conduct granular analysis to determine why it wasn't carrying its weight, Girsch said.

"If you weren't doing 15 percent pretax, you were being looked into in much more depth. We didn't like that," Girsch recalled. "There might be reasons for it, but we were trying to figure out how we could manage through those issues."

The corporate team would frequently fly in to take a closer look, visits not particularly welcomed by local managers. If they were tipped off about their seniors' arrivals, some local managers might race to clean up the shop, prepare, and take steps to address and anticipate the coming interrogation. Safety stripes, freshly painted in maintenance areas, might not have even dried. Good housekeeping said a lot about quality management. Once there, the corporate teams were good at identifying problems.

As they flew in for local reviews, the corporates would joke that "we're here from Oak Brook and we're here to help," Girsch recalled facetiously. The targets of those visits might smile, but their souls were thinking, "Oh [expletive], we don't need these guys here."

Girsch also established the internal audit function, another deeper, equally serious level of review. He assigned Mike Cole to manage it and, later on, Paul Pyrcik assisted. "That was a bit hard to get local operations managers to embrace," Girsch recalled. But it wasn't window dressing. The Policies and Procedures Manual (PAD) was the law according to Oak Brook. "It was taken very, very, very seriously," recalled Dennis Grimm, who ran the eastern region. No one in operations wanted the worrisome blots on one's record that the internal audit might turn up.

Bad audits meant trouble for the managers: Changes and quick action would likely follow a troubling report. "If it was something serious, you had big corrective measures to take," Grimm said. "Nobody wanted to get a note back from Phil that Dean wasn't pleased." It was a prod to always improve. "We gave Girsch all the credit in the world for that," said Grimm, who, after leaving the company years later, would take the MOR process to his other ventures.

"At the local level, it was more of an operations management issue," Girsch said. "The controller didn't have the last say in operations. We didn't give poor managers a lot of string. We made changes. We would try to work with the guys, try to develop their talents, give them time and support. If they weren't able to adjust and make changes and decisions and improve things, they were not going to be around for long."

THE REGIONAL CONTROLLERS

In time, Girsch would have other senior assignments in operations and finance in both the solid and hazardous waste businesses. He'd assist the international team, too. The company would add more financial firepower when Jack Cull and later Tom Hau, both Andersen partners, served as company controllers.

The regional operating and finance teams worked together closely but reported up the chain separately. While there was a wall between them, there was cooperation and communication, too. They had to be able to achieve optimal performance.

In the Southeast, which mainly encompassed Florida, regional manager Eberlin and his successors were supported by controller Jerry Veach. In the Midwest, Jim DeBoer had Bob Brach. Manager Bill Katzman in the northern region had Dick Ancelet. Out west, manager Harold Smith was followed by Dave Pearre, who was supported by controller Tom Collins. Jay Rooney, Phil's brother, looked after the Northeast. His controller, another Andersen alum, was Jeff Lawrence.

Over time, the names and the geographies they directed might change. There would be shuffles and newcomers and regional adjustments. The company's bench strength was deep. Industry veteran Rich

Evenhouse oversaw the Great Lakes region for a time and then the East. Joe Jack, a Saudi veteran, was in Florida. Fred Roberts managed Midwest Disposal in the early 80s. Others would join.

Morrow guided a region comprising the eastern, southern, and central states that stretched from the Carolinas to Texas. Jerry Veach did double duty there for a while as controller until Doug Allman came in.

Morrow impressed everyone. "He was the granddaddy of statistical analysis on how to run a garbage company," Girsch said. "He had books on each and every division. He was a tough guy to work for. He demanded excellence, and he wouldn't accept anything but. If you weren't producing at a high level, you were on his list and it wasn't a good list to be on." Sadly, Morrow died suddenly after suffering a heart attack while golfing in 1985. Across the country, his colleagues mourned.

BIG AND SMALL ACQUISITIONS CONTINUE

From the company's beginning, the energetic Rooney had displayed his managerial talents and business sense. He had easily moved from being Buntrock's assistant to taking on key roles in operations and management. His leadership continued to be recognized, and he rose to the top. He knew the business from the ground up, and he knew where he wanted to take it. He had a talent for dealing with people, motivating them. He was coach, counselor, and teacher, roles he once said he might have pursued had he not joined Waste Management as a young man. Rooney played a leadership role in driving the company's growth and success. "Phil would sometimes just demand that we get into a new market," Girsch said. "'Why are we here? Why aren't we there?'" Rooney would ask. He had high goals and elevated standards. The regions would respond. His people understood. They would deliver.

Year after year, beginning in the early 1970s and into the 90s, the acquisitions kept coming. The largest in the formative years was the $423-million merger with SCA Services in 1984, the nation's third-largest garbage company. It was won in tandem with a Canadian company, Genstar Inc., in competition with BFI. Because of antitrust concerns, the company first reached a novel agreement with the Justice Department hatched by Flynn. The agreement required the company to simultaneously close the acquisition and unload assets to Genstar in certain markets where, in the government's view, the merger would create too great a market presence.

The SCA deal was the biggest so far, and it catapulted Waste Management ahead of industry leader BFI in size for the first time. SCA's assets broadly expanded Waste Management's footprint to forty-three new markets, such as Little Rock, Mobile, Oklahoma City, Memphis, the Portlands in Maine and Oregon, Cleveland, Louisville, and Washington, DC.

There were always the "tuck-in" acquisitions, too, where a few routes would be added, generating greater density, productivity, and, therefore, profitability on collection routes. At the end of 1984, the company had gained one hundred and fifty new communities. The number of household customers had grown to nearly 4.7 million from 3.1 million just a year earlier. Its business had enjoyed enormous growth: Revenues now reached $1.3 billion.

Huizenga, the co-founder with the persuasive blue eyes, had gradually reduced his role and decided to retire that year, the first in which the company reported revenues exceeding $1 billion. Before the 1971 public offering, he had promised Buntrock he would stay five years. He had more than doubled that commitment. He was credited with leading the acquisitions program that expanded the company's North American businesses. He had been Buntrock's partner, and his continuing counsel would be valued.

"Wayne's contribution of strong leadership, boundless energy, clear probing mind, and delightful sense of humor led the company through many historic decisions," the company told stockholders. "Wayne will continue to maintain a close relationship with the company in the future by serving in a consulting capacity."

Huizenga would not rest; he had other interests, many other interests, and certainly enough to keep a retiree busy. He would gain greater fame and fortune in other businesses. He invested in hotels in Fort Lauderdale. He became known for growing the Blockbuster video chain, which began with just eight video stores and would eventually go international with 3,700 locations in eleven countries before being sold to Viacom for $8.4 billion. He had an interest in the Extended Stay America hotel chain. He created the AutoNation chain of car dealerships and took it public. Later on, he reentered the garbage business with Republic Industries, which would become the second-largest waste handler in North America. And he would own three professional sports franchises in South Florida: the NFL's Miami Dolphins football franchise, Major League Baseball's Miami Marlins; and the National Hockey League's Florida Panthers. There was a lot on his plate.

Meanwhile, Waste Management was still growing, still acquiring. Thanks to the federal regulations they had foreseen years earlier, the hazardous and low-level nuclear wastes businesses showed promise and profitability. They had acquired resources in this market early on. It would add profits; it would pose problems, too.

Throughout the 1970s and 80s, the executive team could not have been busier domestically, but unusual and fortuitous opportunities arose overseas, taking the US-based company global.

TEAM MEMBERS LISTED IN THIS CHAPTER

Ron Baker	Jim McGrath
Larry Beck	James Mears
John Blew	John Melk
Dave Blomberg	Guy Molinaro
Dean Buntrock	Bob Paul
Jerry Caudle	Mike Popovich
Wally Carlson	Don Price
Curtis Collins	Gene Price
Mike Curry	Jerry Rhodes
Rich Evenhouse	Mike Rogan
Ed Falkman	Phil Rooney
Don Flynn	Roland Schmidt
Tom Frank	Lou Staniszewski
Clair Hoeksema	Joe Szyluk
Dale Hoekstra	Bill Thon
Peter Huizenga	Wayne Thurman
Wayne Huizenga	Fred Weinert
Rod Jarvis	Martin Weller
Jackie Jacobie	Duane Western
Bob Keleher	Mick Witnell
Jim Koenig	Bob Zralek

11

THE ROAD TO RIYADH

It was a challenge like no other the young company had yet faced. And it could not have come at a better—or some believe, more critical—time.

Waste Management, only a handful of years beyond its 1971 public offering, was struggling along with the rest of the nation in the mid-1970s. A recession that had begun late in the Nixon White House years had lasted into the Ford years in 1975. Bad economic news was consuming barrels of black ink in the headlines: The 1973 oil crisis, followed by a two-year bear market on Wall Street and other global stock exchanges, caused share prices to plummet. Unemployment increased and right along with it, inflation. A new term was even coined for the phenomenon: "stagflation."

Waste Management's stock, the currency it routinely used to attract so many modest and family-owned garbage companies into its fold, had dropped considerably. In 1975, it fell to five dollars a share (far below its 1971 initial public offering price of $16 a share), a price no longer as enticing to acquisition prospects and no longer useful to the company's managers, who understood there was no value in diluting their anemic shares to snap up acquisitions.

But against this unfortunate economic backdrop, opportunity arose.

The problem was that it was 7,000 miles from Chicago in the middle of a Middle Eastern desert.

A PROSPECT IN RIYADH

In 1975, Buntrock saw an article in a European newspaper about the Kingdom of Saudi Arabia seeking bids for sanitation services in its capital city of Riyadh. Saudi Arabia was flush with cash and eager to spend lots of it after the 1938 discovery of oil and ongoing development of seemingly boundless reserves of it beneath the sand along its Persian Gulf coast.

The Saudis were spending billions to upgrade Riyadh's infrastructure. In particular, they were searching for a sanitation company to bring Riyadh, a city with about 700,000 residents, into the twentieth century by installing a capability it did not itself possess. They wanted to rocket Riyadh from a traditional society into a modern state. Riyadh needed what was essentially a municipal public works department to sanitize the place. Namely, solid and commercial waste collection (the Saudis had little to no industrial waste); transfer and disposal; street sweeping; dead animal cleanup; and abandoned car removal. (With few towing services, Riyadh's drivers left their broken-down Mercedes, Chevys, and other vehicles along the roadsides.) Its city government hired a British engineering firm to assist them.

At first, the dollars in the tender seemed too good to be true, too enormous not to look into. The project was worth $500 million over five years in 1975 dollars (about $2.35 billion today), with a big chunk of cash delivered up front. No one at Waste Management really believed the high-end number, and eventually, it would descend to a more down-to-earth figure. But still, the offering seemed plenty rich enough for Buntrock.

But how to earn the cash? Buntrock was curious to investigate the tender. The numbers were just too huge to ignore. He wanted to take a look. He got little encouragement from his fellow officers—they were, let's say, skeptical. Peter Huizenga favored the move. Wayne Huizenga and Beck were not interested. After all, they had their hands full conducting the difficult job of rolling up new companies domestically. But they didn't push back on the idea either. The tenacious CEO kept persisting and pursuing, displaying his quiet but determined South Dakotan mien.

"Why don't you take a ride over there and see what it's all about," Buntrock told John Melk and Don Flynn, as though they were hopping on the next commuter train from suburban Oak Brook to Chicago's Loop. Exiting the room, the two had their doubts. They weren't even sure about the country's geography. "Where the hell is Saudi Arabia?" Melk jokingly recalled Flynn deadpanning.

The two soon arrived in Jeddah's two-Quonset-hut airport on their way for their first "look-see," as Melk recalled it. They had missed their connecting flight to Riyadh. The stocky Flynn was uncomfortable, sitting and sweating on the cement floor with goats wandering nearby. It was 120 degrees and humid, Melk said, and they wanted tickets to anywhere else. Luckily, Melk saw an opportunity to snatch two first-class tickets to Riyadh.

They were on their way. The Riyadh investigation was launched. "At that point, we were pretty fresh-faced kids," Melk said. "We were really starting from scratch."

HIGH AMBITION PAYS OFF

After the first visit, Melk dove into a year of learning what was needed for the bid, understanding how to respond to the Saudi culture and government, figuring out the decision-making process, making contacts, and gathering resources. The company needed to know how to present itself, what was needed to build relationships, and the who and how of the final contract award. And they needed to field a team that understood the scale of the project and the hard work of submitting a bid. Melk was concerned that their competitors had greater influence, and it appeared they did.

The project progressed in fits and starts. At one point it would grind to a halt. Melk and his team had to navigate different levels of Saudi government—municipal, provincial, and national. Without connections, getting a meeting with the right people and getting them focused was hard. Melk decided to write a letter to the Crown Prince. The three-page appeal worked. It seemed to open doors. "After my letter, all of a sudden it changed," Melk said. "They welcomed me, had meetings."

The company's leaders had zero experience operating overseas. Even so, Melk's team was unafraid of overreaching. They simply didn't know what they didn't know. Waste Management may have been young and small, but its team did not lack ambition. The twenty- and thirty-something leaders had abundant energy to pursue big lofty endeavors.

While Waste Management had been rapidly expanding in its earliest years across the US, it was still barely removed from its humble beginnings in Chicago, Florida, and Milwaukee. And it would be up against formidable international competitors who had experience in the Middle East. While Waste Management had been quick and facile in acquiring and integrating garbage companies into its operations, the Saudi project would require a huge multinational mobilization across the globe. Ambition did not equate to execution.

THE RACE TO WIN THE BID

The Melk team knew they couldn't go it alone. They formed a joint venture with Pritchard Services, a British provider of airport support and cleaning services, to chase the project. Pritchard, led by P.R. (Peter) Pritchard, knew the Middle East better than Waste Management and had been trying to get work there. The company agreed to Waste Management owning the majority interest at 60 percent and Pritchard, 40 percent.

Waste Management also needed a local partner. They found him in Prince Abdul Rahman bin Abdullah bin Abdul Rahman, whom Pritchard included in its share. The prince was a member of the royal family and gave the venture a very helpful, valuable, conservative, and credible connection.

Waste Management did not originally know the prince, but "the prince had heard we had an interest," said Buntrock. "After we had been there [investigating the tender], our prince called his partner

[Pritchard] in the UK and said he'd consider being a partner if they got the American company. He selected us. We could never have been involved without that."

The prince had family, connections, and clout. Lots of clout. He was the nephew of King Abdulaziz, who, after a series of conquests, established the Kingdom of Saudi Arabia in 1930. The prince's father, Prince Abdullah, helped his brother establish the Kingdom and was rewarded with large tracts of land, believed to have made the family the largest landowners in Saudi Arabia, according to Ed Falkman, a lawyer who was recruited in early 1977 to be general counsel for the Waste Management partnership.

The conservative, play-by-the-rules Prince Abdul Rahman served a key role. "He was somebody nobody wanted to fool with," Melk said. "It took people who were afraid of him out of it. And we weren't willing to do any of the monkey business with payoffs. Ultimately, it worked fine."

The partnership's powerful and more experienced international competitors were Générale des Eaux, the big French water and waste management company (today Veolia Environnement S.A.); Plastic Omnium, a maker of wheeled bins; and Groupe Renault, which built trucks. The latter's agent was the brother-in-law of Riyadh's then-mayor. "It was a project customer-fit for the French consortium," Falkman said.

Beginning in 1975, Waste Management revised its bid documents three times, a process that took two years to complete. The first bid was for door-to-door collection twice a day, and the second reduced the pickups to once a day. The partnership was the low bidder on both. "So that presented a problem for the French consortium, which had been instrumental in getting the project conceived," Falkman said. "By the time it got to the third bid, the specs were four volumes."

A British engineering firm helped prepare the project specs for Riyadh. "It was an absolute hodgepodge, cut-and-paste," Falkman said of the specs. "For every clause in there that said one thing you could find a clause somewhere else that said just the opposite. But the gist of it was they wanted a contractor to be responsible to collect all the waste in the city of Riyadh and be responsible for all the growth in the city as it grew. And we again were low bid on that third bid." The engineering firm included a map that marked the city's boundaries, which would prove critical in the months ahead.

Sheikh Abdullah (more formally known as Abdullah Ali Al-Nuaim) became Riyadh's mayor after the second bidding. By the third bid, the deputy minister of the Ministry of Rural and Municipal Affairs, whom Melk had wisely developed mutual respect with, began overseeing this project.

Falkman remembered well the final meeting with the ministry in July of 1976. "The bids were opened and the partnership again was the low bidder. The minister asked the French representative if it was his last and lowest bid. He said no and offered a price that was a bit lower than that offered by the Waste Management partnership. So he turned to John Melk and said, 'Is that your last price?' And John being clever, said, 'Before I answer that, I would like the French representative to say whether or not this is his last and final price,' to which he responded yes. And John then said, 'Okay, in that case, my price is blah, blah, blah,' a little bit lower, and we were low bid and awarded the contract."

Years later, the Waste Management team learned that the French, who had a large team working on the bid, had a communications problem because the bid opening was on a Sunday, a day when French families got together and company leaders were not always available. The head of the French partnership had been out of reach. Even if he had not been, Saudi telephone connections could be problematic and often had to be arranged days ahead.

Melk described the win as "a monumental opportunity we should probably never have gotten. We were starting from scratch. In addition to managing wastes, there was street sweeping, abandoned car removal, and stray dog collection. Every carpetbagger was going there trying to get some of that money," he said.

Mike Rogan, who became the project's finance director, agreed. "The audacity to think you could even get involved with something like that is quite striking," he said. "This was going to involve activities that Waste Management never had any experience in. We'd never built a housing facility. We'd never shipped trucks from overseas, we'd never hired foreign workers. Such a mobilization would have scared away just about everybody."

FINALIZING THE DEAL

But winning the tender was just the start. They still had to negotiate a contract. Negotiating sessions would go on and on. Days would pass, then weeks. Little progress was made. Waste Management, with its growing experience in acquisitions, knew it needed legal help.

Peter Huizenga, a lawyer and Waste Management's corporate secretary, was designated in charge of legal. He hired a Saudi lawyer, Salah Hejailan, during one of their early trips to Riyadh. Hejailan was one of a small number of Saudi Arabian attorneys assisting Western clients, and Waste Management's team met with him to "learn how to do business in Saudi," as Melk described it.

For their first meeting with Hejailan, the team climbed up a winding staircase to his fourth-floor office as the building's elevator was inoperable, Melk recalls. They stepped over surrounding debris to reach his door. Accompanied by Melk and Flynn, Huizenga sat in the middle of three seats in front of a large desk. After a long wait, a Saudi dressed in robes came in, sat at the desk, took out a pack of cigarettes and smoked one, Melk recalled. After some time, one of them nudged Peter and said, "You're in charge of legal, say something!" Huizenga launched into his introduction on why they were there while the man lit another cigarette and stared silently back at them. Finally, the office door opened and the Saudi behind the desk got up and left, replaced by a man dressed in a flowing robe decorated with gold. "'Hello, I'm Salah Hejailan, how can I help you?'" Melk recalled him saying. "Peter had been talking to the tea server."

Huizenga also hired John Blew, a securities lawyer in the Chicago law firm of Bell, Boyd & Lloyd, who became Waste Management's principal outside counsel. He had gained some Saudi Arabian experience working on an infrastructure project to electrify the country in the early 1970s.

Blew was asked to review the tender and help prepare a responsive contract. He dove in. His planned two months of duty turned into six and ultimately stretched into two years of back-and-forth trips.

"Waste did not have fixed office space. People would work out of their hotel rooms or a conference room in a hotel," Blew said. Other Waste Management team members lived in a bed-and-breakfast. Blew set up his legal shop in his two-room hotel room. It was not luxurious, with just a bedroom and a living room, and the accommodations were far from the niceties of Riyadh's first-class Intercontinental Hotel. Still, Blew's modest living room offered enough space for the team to meet, spread out papers, and review materials. It served the purpose.

Working through the contract process was challenging. The Saudis had engaged Watson Saudi Arabia, an English engineering firm, to help prepare a request for proposal (RFP), and they set a tight deadline.

The RFP was prepared in English and Arabic and had an English law perspective. It was about an inch thick. Some of the contract language was difficult to understand and Blew worked to identify issues where the team needed to qualify its responses to contract terms. There would be a stream of revisions. The initial contract drew comments from the Ministry of Rural and Municipal Affairs' advisors. Staff in a mobilization office in Oak Brook helped to interpret the ministry's questions and guide the Riyadh team in responding.

Handling the edits wasn't easy. The contract kept changing, and Blew would have to then incorporate the revisions. "We'd work on this contract and get it all typed up in Riyadh, which was hard to do, and change it over from Arabic to English. He'd have to get on an airplane, fly back to Chicago, get it in a contract form and fly back," Melk said.

Blew's expertise was critical to the effort, Falkman said. "He was one of the most phenomenal draftsmen I've ever come across. He would draft pages on his legal-size yellow notepads. He just had a wonderful ability to draft contractual terms in a logical, clear, and concise manner that could easily be understood." At one point, an exhausted Blew needed a break and asked to go home. Melk and Flynn wouldn't allow it. "John got instructions that he wasn't going anywhere. He needed to stay the course," Falkman said.

The Saudi contract law was based on British standards. "It was just very informal by our stateside standards at the time, and basic," Blew said. "Waste Management didn't under-lawyer anything. They lawyered up early, and the Saudi project was no exception."

Blew also had to adjust to the culture and different norms. Early on, Blew hired an American woman to help with the drafting and revising. Each floor of his hotel had a guard from the religious police stationed down the hall. When the woman knocked on Blew's door one evening, the guard was immediately on alert. "She wasn't in the room more than a minute and there was another knock on the door, and it was this guard down the hall who wanted to know who she was and why she was there," Blew said. "It took an hour to straighten it out. It was typical of the things you found yourself in. None of us knew what this incredibly alien culture was all about."

Arranging negotiating sessions with the Saudi ministry could be agonizing for the Waste Management team and different from the company's style. Melk recalls one meeting with the minister, his deputy, and five or six others. Melk and Blew were accompanied by several colleagues. The meeting lasted for a couple of hours, and then the minister got up and excused himself, never to return. Then the deputy minister did the same, and soon, all of the Saudis left. "Well, I guess the meeting's over," Melk remembers thinking. "That's the way it was. You wondered what you accomplished and when you'd have another meeting."

Another time, a day before an American holiday, the Riyadh officials asked to meet with the Waste Management team. Flynn and Melk left their families in the US for the holiday and flew to Riyadh. No one met them for days.

The final documents for the five-year contract called for building a complex of eighty-eight buildings to house and feed four thousand workers, maintaining one thousand vehicles, producing power, and treating water and sewage for the complex, Falkman said. It also called for Waste Management to recruit, train, and manage an expatriate workforce in the thousands and import one thousand vehicles at a time when there was a six-month delay at the ports. The contract spelled out where the laborers would be housed and how they would source the labor, what kind of equipment would be used, and how many trucks and what type of trucks would be required. The details went on and on.

"There were not standard trucks for the narrow unpaved streets in Riyadh at that time," Blew said. "There were narrow lanes with high mud walls on each side. Sometimes you could almost reach across them. The Waste Management partnership had to adapt a lot of equipment, and they had to specify how those vehicles would be able to perform the work under those conditions. The Saudis were not able to directly monitor this, but their English consultants could."

Waste Management had to complete these measures within 360 days of the contract, at a fixed price.

Finally, after six months of negotiations, the contract between the Waste Management-Saudi Pritchard Joint Venture and the Saudi Arabia Ministry of Municipal and Rural Affairs was signed on January 31, 1977. Its value was $243 million over five years, about half of the original figure that had caught Buntrock's attention, but still plenty large enough in Waste Management's eyes.

The contract value nearly equaled Waste Management's then-annual revenue of nearly $250 million, about $1 billion in 2020 dollars. With the signing came a 20 percent down payment of $50 million, a welcome and much-needed infusion of cash to start the mobilization, buy equipment, share with partner Pritchard, and add to the company's treasury.

CHASING THE MONEY

Getting that first payment and subsequent payments, which were tied to project milestones, was not always easy. Waste Management had to navigate the Saudi process. First, the partnership would create an invoice to take to a Riyadh financial functionary for approval and a signature or stamp. Next, it would be taken to a Saudi official at a higher level to approve. Then it needed approvals from the ministry and, finally, the Saudi Arabia Monetary Authority. Only then could the check go to the bank for payment.

"There were a lot of steps and a lot of places I had to go," Rogan said. "Typically, the person you would meet with would say, 'Have a seat.' And you'd sit there and you'd drink tea. He may be doing business with someone else, and you'd sit there. Then he'd say, 'I have to go, you come back tomorrow.' And you'd come back tomorrow, and the person might not show up for two or three hours. Over time, it took almost forty-five days to get the collection with me going from office to office with no action and sometimes being asked to get something from the contract that would cause us to go backward. Now the only reason I kept doing that was because I knew at the end of the day, I'd be asked by Bob Paul [the finance director] or Don Flynn [vice president of finance] 'What's the status of the money?'"

To receive the initial down payment, Rogan, accompanied by P.R. Pritchard, went to a Citibank location in Riyadh, where he learned the payment was actually being held in a Citibank branch in Bahrain. "The next day Don Flynn comes in. He got on a plane right away," Rogan said. "We were going to wire transfer it. Don and I went to Citibank to arrange for the transfer. Don Flynn asked him how many dollars we were going to get." The Bahrain Citibank manager gave them a final rate and the details. But Rogan, having read a news story on exchange rates, questioned the rate. "Don asked, 'Why did you ask him about that rate? I think you just saved us $15,000 just by challenging the rate.' We were learning on the fly."

Rogan not only chased the payments, he ran the finances. He later installed up-to-date accounting systems and created a policies and procedures manual for the site's workers.

Flynn made many trips to Riyadh. He hated the long travel, Buntrock remembered. "We came back from Saudi on Wednesday. Thanksgiving was Thursday. I got a call from the prince. He wanted us to come back for a meeting. I remember calling Flynn that night because it was financially related. I told him we had to go back to Saudi the next day. He gave me sixteen reasons why he couldn't. But he was on the plane."

There was constant back-and-forth with Oak Brook. "Buntrock thought nothing of being on the phone for hours," Blew said. "Phone connections were not easy then. He was so involved. He understood everything. He went over it with a fine-tooth comb." On one occasion, Buntrock was in Oak Brook poring over some final contract deals and called Flynn in Saudi Arabia to discuss them. Buntrock soon heard snoring. The exhausted Flynn had fallen asleep.

MAPPING THE STREETS OF RIYADH

As soon as the contract was signed, Melk organized a larger team to plan the project's mobilization. An in-country crew communicated with the Saudis. Back in the States, Senior Vice President Phil Rooney dispatched a survey team to try to estimate the volume of waste to be managed and the other resources needed.

The team included Don Price as assistant team director and a respected roll-up-your-sleeves, do-everything manager from Chicago; his brother Gene Price, who had worked in acquisitions; and Ron Baker, an operations manager from Holland, Michigan. Baker had never had a passport and needed a letter from his minister proving he was a gentile before he could go to Saudi Arabia. The team also included experienced company operations managers Wally Carlson and Duane Western from Milwaukee, along with Martin Rivers and B. Workman, the latter two Pritchard men from London. The majority were experienced hands at acquiring and integrating collection companies in the US, counting trucks, charting routes, and collecting waste. This project was different.

No one had a clue how much waste there was; there had never been a system to collect or count it.

The teams scoured Riyadh, Price said. They walked across the city, taking notes and photos and making calculations on the kind of waste volumes that would likely be generated and how they could organize routes to collect them. Further complicating the counting and planning was the fact that the city had never been mapped. To create collection routes, the team would have to develop Riyadh's first city maps.

"We divided the city and walked the streets," Price said. "There were goats everywhere." They sprayed the goats with different colors in an effort to tie them to households. "That was to figure out how many containers you would need, he said." Back in the States, there was plenty of data available on waste generation. Riyadh was another world.

As hard as the team tried, their waste estimates were often more guesswork than clear calculation. The Saudis were focused on cleanliness, and so an accurate waste volume estimate was critical: It would determine how many workers, collection trucks, support vehicles, routes, transfer sites, and other support were needed.

"Whatever it [the waste volume estimate] was, it was hugely underestimated," Falkman said. "It was seriously below what was required. There really was nothing you could turn to. There was no collection system in place. There were some private contractors who might have two trucks that went around and collected some things. All the houses were inside a walled compound, and they would put the waste on the outside of the wall and the goats would eat the waste. There were goats wandering everywhere. And then there was all this rubble everywhere, big concrete chunks, cement blocks, lots of stuff, asphalt."

THE MOBILIZATION STARTS

Meanwhile, the mobilization began in earnest. Phil Rooney served as director of mobilization, a "godawful job," one executive recalled. He brought leadership, operations savvy, and organization to the task. Dave Blomberg was named the site's project manager. Rich Evenhouse and Jerry Caudle, experienced veterans in US waste operations, joined. Rogan moved out of Flynn's accounting group and directed financial matters. Most of the team was still in their twenties and early thirties.

Flynn controlled the money. He liked having cash in the bank. He had negotiated favorable, 120-day terms with equipment vendors for payment. The challenge, said Bob Keleher, who was responsible for construction scheduling, was to move the equipment from the manufacturer to a port for shipment to the Kingdom and have it accepted before the Saudis would pay. Only then would payment go to vendors. Flynn's plan allowed the company to use the initial upfront advance payment to fund expansion in the US, including acquiring assets to manage chemical wastes, a then promising opportunity for the company.

The purchasing team in Oak Brook, led by Rod Jarvis and including Bill Thon and Lou Staniszewski, was buying shiploads of trucks and heavy equipment and provisions, most of it purchased domestically, to construct the desert facility. Joe Szyluk, a retired Aramco executive, joined the team to guide and coordinate the enormous purchasing activity. The project called for rear-loader collection vehicles, street sweepers, load-luggers, stationary compactors, pickup trucks, passenger cars, 3,000- and 5,000-gallon water tankers, a fire engine, landfill compactors, scrapers, bulldozers, residential and commercial refuse containers along with materials to maintain them. It was a long list. International Harvester supplied the trucks, Heil the collection bodies. To clean the streets, they brought in Elgin Sweepers. Terex provided dozers, scrapers, and other heavy equipment for the landfill, and Massey-Ferguson the lighter mechanized tractors used at transfer stations.

Shipping the equipment from the US to Saudi Arabia was another challenge. Waste Management drivers in the US ferried the equipment from manufacturers to ports in Baltimore and Norfolk and then flew back to the plant to drive yet more equipment to the ports.

There was also a six-month delay in unloading at Saudi ports right when the team was shipping its equipment, housing, and supplies for the start-up. The team used ships designed for roll-on, roll-off cargoes to avoid delays.

THE CAMP IN THE DESERT

Before the team could even think about cleaning the city, they needed to establish an operations center. The project called for constructing a completely self-contained community to house and support 2,000 employees, manage operations, maintain and dispatch a 1,000-vehicle fleet, distribute 500,000 disposal bins, and provide disposal capability. The mobilization team called it the Camp. It was more like a small city. The project required the construction of thirty-one transfer stations across Riyadh where waste would be dropped for transport to the landfill. Everything had to be imported. There were no Home Depots down the street.

The project figures were staggering, the timeline tight—just 360 days to build it, per the contract.

The Camp was located in the desert ten kilometers outside Riyadh. Fred Weinert was put in charge of building the thirty-two-acre complex. There was nothing there, only cinder blocks marking the site. It would be a massive undertaking.

Before Weinert began this assignment, Buntrock had asked him to look over some of the work papers with cost and pricing reviews for the construction project. Weinert recalled Buntrock saying, "'Please review it and let me know what you see.' He didn't say much else about it." A few weeks later, Weinert attended a dinner in Oak Brook, where he sat next to Rooney. "He told me, 'Fred, we want you to be the director of construction on the Saudi project.' I would have scratched my head," Weinert recalled. "I told him, 'Phil, you know that I don't know shit about construction.' He said, 'No, no, we know that. We need somebody to manage it. We need somebody who knows the details and has been through the reviews.'" With that, the job was filled.

Weinert was soon in Riyadh driving the senior Waste Management team, including Buntrock, Huizenga, Flynn, and Melk out to the Camp's location. The five were squeezed into a white Chevy Impala when the car slid off the desert road. Its back wheel got stuck in a culvert. Weinert was horrified and embarrassed. He and his bosses were forced to push the car back onto the road. They thought they were in the middle of nowhere—no water, no electricity, no nothing, Weinert said.

Weinert hired Bechtel Corporation, a San Francisco engineering, construction, and project management company, to design the facilities. He chose J.A. Jones Construction of Charlotte, North Carolina, as the general contractor to construct the Camp and landfill, because they were attentive to preserving the budget, though the partnership team made the most of the construction decisions. Jones later brought on a South Korean construction firm to help with construction, which added much value to the project, Weinert said. "The last thing you want to do is pour concrete in the middle of 110-degree weather. The Korean construction team figured out you would do it at daybreak, and it was a good decision."

Waste Management then began to deploy its own team to oversee the design and construction. Bob Zralek, a retired rear admiral with thirty-seven years of service in the US Navy, had just retired from the City of Chicago as deputy commissioner of streets and sanitation and joined the group in London. A former commander of the Glenview Naval Air Station near Chicago, he worked with Bechtel to cook up a design for the site that closely resembled a US naval air station. Zralek found more talent: Mike Popovich, a retired Navy Seabee, as construction manager; Mike Curry, a mechanical engineer; and Jerry Rhodes, nicknamed the "Rhodes Scholar," for overseeing maintenance facilities. Perhaps unsurprisingly, some recruits shared Zralek's Navy experience.

Initially, construction purchasing and the acquisition of housing were run out of the Pritchard Services' London office and later, at a new Waste Management location there, where Jackie Jacobie became a key administrator. Kitchen and laundry equipment, small refuse containers, and furniture and fixtures were acquired through London. Johnson street sweepers were bought in the US.

A couple of months into the mobilization project, during a Saturday lunch in Oak Brook, Weinert recalled Flynn asking, "What do you need help on?" "I just need somebody who can think," Weinert replied. "I know somebody like that," Flynn said. The guy was Jim Koenig, another

Arthur Andersen alum. "Jim and I almost became like a tag team," Weinert said. "He became my assistant on construction."

CONSTRUCTING THE CAMP

The partnership team hurried to meet the start-up date. The first concrete was poured on July 7, 1977. Construction included 256,000 square feet of space in eighty-eight buildings, with rooms for housing employees, supervisors and their families, administrative offices, dining halls, recreation, warehousing, and maintenance.

"We didn't have water there, we had to bring it in," Weinert said. "We didn't have sewage, we had to build a sewage treatment facility. We had to make potable water, we had to get electricity. All the things you take for granted had to be put there."

The team found good vendors. Construction purchases were primarily sourced in Europe. "Most important was Block Watne," Weinert said. "They were the largest housing builder in Scandinavia and the pioneer in manufacturing prefab housing in Norway. They provided all the prefab housing initially for the two thousand workers and provided the mess halls, ablutions, laundry, commissary, and clinic. It was manufactured in their factories in Stavanger, Norway, and shipped to Riyadh in forty-foot containers."

The schedulers were under a tight timeline. Like other Scandinavians, many Norwegians were on holiday in July, and Bechtel needed to get as many prefab buildings into shipping containers as possible before vacations started. The buildings, interestingly, were designed and insulated for Northern Europe's cold temperatures, which also helped ward off the Saudi heat. But the doorways had raised thresholds to keep out blowing snow, not a likely prospect in the desert.

There was separate housing for the expatriate managers and their families. The first housing for expatriates included rental units in Riyadh. People moved to the Camp as construction made more housing available. A number of Americans brought their families, including Rogan, Caudle, Blomberg, and several Pritchard employees. Occupancy was determined by position and responsibility.

Six dining halls totaled more than 23,000 square feet. Mess halls were divided into two areas, allowing for one to be shut down and cleaned while the other remained open. Food service personnel prepared more than 40,000 meals a week. Kitchens had walk-in cool rooms for food storage and preparation, as well as dishwashing and storage areas. A separate 7,750-square-foot mess hall served management and expatriate supervisory staff.

The Camp's recreational facilities included two twenty-five-meter swimming pools, tennis courts, a squash court, softball diamond, soccer field, cricket pitch, and sitting and reading rooms. It also had a 6,500-square-foot clinic and commissary with two doctors and five nurses. Employees could access a commissary for personal needs. There was a full-service laundry.

The Camp also addressed the Muslim employees' spiritual needs. The team built a mosque outside a corner of the site's walls where workers and others in the area could worship.

Entirely self-supporting, the Camp had a clean water system that provided 120,000 gallons of potable water, along with raw water and treated water storage tanks. It included wastewater treatment and irrigation systems. Power plants provided 4,800 kilowatts of electricity to the main complex and a separate system served the landfill and incinerator sites. They also created a telephone system with 124 stations, an intercom system, and an FM sound system.

The Camp's construction was completed on March 31, 1978, in just nine months. They had beaten the 360-day deadline.

RECRUITING AN INTERNATIONAL TEAM

Of course, the point of all this construction was to clean the city. And to do so, the project needed workers. The hunt was on to find them. Tom Frank, who had joined the company in its earliest days as Buntrock's assistant, was enlisted to recruit a team and find and draft the army of workers needed.

Frank moved his family to London but ended up spending months in Saudi Arabia and India, where he recruited many employees. In Britain, he led the recruitment of a hundred or so experienced managers and supervisors from municipal governments, where they had been performing waste management duties there.

Initially, Frank and Rooney were looking east from Saudi Arabia to recruit employees in Pakistan, with its population of 70 million. They were in final negotiations when Zulfikar Ali Bhutto, the country's then-prime minister, suddenly doubled worker wages. (Bhutto would later be deposed in a military coup and subsequently executed in 1977.) "Our hearts sank," Frank said. With Bhutto playing hardball, Frank and Rooney consulted with Oak Brook headquarters. "Pack your bags, you're going to India," Buntrock said.

The team then focused their recruitment efforts on India and engaged MacKinnon, MacKenzie & Co., an agency that mostly supplied workers to the shipping industry. James Mears, a retired labor relations officer from Aramco, was brought in to help. Mears was fluent in Arabic and had helped write Saudi labor laws. He had also lived in the Middle East for years. "We hired him early on to give us seasoned advice," Frank said. "He was very articulate and Middle East savvy. We needed someone who could guide us as we tried to sift through the candidates."

Frank's recruitment group set up shop in the MacKinnon, MacKenzie office in Bombay, now Mumbai. Frank brought in Jim McGrath, an Oak Brook human resources manager, and a number of other Waste Management employees. Nearly a dozen members of this team spent thirteen months living in the luxurious Taj Mahal Palace hotel (though their stay was made less luxurious due to frequent stomach upsets from the food and noxious odors from the chemicals used in the hotel's pest control efforts, McGrath said). Among them were Clair Hoeksema, Roland Schmidt, and Guy Molinaro, who assisted with the recruiting, interviewing, and hiring process. They processed candidates and sent them along to Riyadh.

To attract applicants, the team placed a help wanted ad in local newspapers. It brought an avalanche of recruits. "We had literally thousands of people coming in over a short period of time. We had people around the corners, down on the streets, a couple of blocks back waiting to get in to have interviews with us," McGrath said. "They'd be climbing over the eight-foot fences at MacKenzie trying to get in line. The lines were tremendous," Frank said.

Dave Blomberg, who became the resident manager in Riyadh, in effect the city's sanitation commissioner, estimated that more than thirty Waste Management employees were dispatched to Bombay to test and recruit workers. "We were tough. If a man said he was a driver, we made him show us," Blomberg told *Washington Post* writer Thomas W. Lippman in a January 1978 article headlined, "Saudis Hire Americans, Spend Millions to Clean Up Capital." They obtained a bus from a nearby school to test the candidates' driving skills.

Not every candidate got a contract. The selection process was disciplined, and recruits had to pass a physical to ensure they could handle the demands of the job. MacKinnon, MacKenzie's Dr. Dadrawala conducted the exams and turned down those unfit for the work.

"His heart and his ethics were in the right place. He probably turned down about seven thousand people," Frank said. "They would come in for their physical, and they wouldn't be eligible because of hernias and other health conditions. It was a mess, but we worked through it." Frank estimated they may have rejected a third or more of the applicants for health reasons.

At the Camp, there was a range of jobs to fill: refuse collectors, street sweepers, drivers, dog catchers, chefs, kitchen servers, and building cleaners. And they needed accountants, administrative staff, generator plant operators, water and sewage treatment plant operators, warehouse staff, security and fire protection crew members, heavy equipment operators, and mechanics. They even needed a doctor and clinicians.

"The deal was that we recruited all over India, but Bombay was the main area they [applicants] would come to for interviews," said McGrath. "Workers had to sign a contract in which they would agree to go to Saudi Arabia at least for two years, and we would pay them a salary and provide all their food, clothing, housing, health care, etc. They were also required as part of their contract to send part of their salary back to India to be deposited in a bank so that they would have money when they finished their contract." The Indian government mandated that 75 percent of their wages be deposited in India. Employees were paid $250 a month (approximately $1,200 in current dollars), with free room and board. Living in the desert, they had little to spend it on.

Some employees made very little income and were earning a higher wage than they had before, McGrath said. "We figured, well, when they go back to India after a couple of years, or maybe if they stay longer, at least they'll have a portion of their wages sitting in a bank account, which will give them opportunities to move ahead in their lives."

ADAPTING TO CULTURAL DIFFERENCES

Communication with the Saudi embassy a hundred miles north of Bombay was critical. Mears, the retired Aramco executive, helped obtain visas and passports and process candidates. "Anybody who would go to Saudi Arabia had to have a visa," McGrath said. "He would go up there and do the schmoozing."

Once hired, the recruits were flown to Riyadh. "As housing was completed, the occupants were being picked up at the airport and cleared through customs and immigration," Keleher said. "One of their first tasks was to assemble their beds before heading to the mess kitchen for their first meal."

Keleher recalled that with employees hired from all over the world, there were sometimes cultural and language challenges. Many didn't speak English—McGrath estimates that his team hired "800 to 1,000 people who had never spoken any English at all."

Many mechanics were unfamiliar with automatic transmissions, air conditioning, and compressed air equipment. The Americans needed to translate the terminology they used in training. "A *boot* was now a *trunk*, a *wing* was now a *fender*, a *windscreen* was now a *windshield*, a *bonnet* was now a *hood*, a *spanner* was now a *wrench*," Keleher said. "It was a challenge for all involved."

THE BIG CLEAN BEGINS

Shortly before the start-up, someone suggested that the site organize a parade. *The Post's* Lippman noted that "Residents cheered when all two hundred yellow trash trucks rolled through the streets in a parade, sporting stickers on the doors saying, 'Help Make Riyadh Clean.'" He further reported that when the trash cans began to disappear, the newspaper *Al Riyadh* observed: "Regrettably, some children, and perhaps some of their elders too, took possession of these bins and hid them in their houses. It is not quite understandable how the citizens expect to maintain the cleanliness of their capital city given such behavior, which only reveals the extent and ignorance and indifference to the municipality's effort to keep the city tidy." The partnership had distributed an estimated 500,000 plastic garbage cans.

Along the parade route, a few of the collection vehicles got lost, "maybe because of traffic or whatever," Falkman recalled. The drivers eventually abandoned the vehicles but found their way back to the Camp well out of town. The drivers had no idea where they had left their trucks and it took a few days to recover them all.

The day came to execute the contract and begin cleaning the city. The mobilization team wasn't quite ready. Waste had piled up in Riyadh for years, and the Saudis expected it to be cleaned up on day two. Employees and equipment were still arriving, but the work had to begin. The partnership's agreement with the Saudis was an at-will contract, meaning the Saudis could terminate it at will on short notice. "It was not a favorable termination provision," Blew said. "Waste was pretty much locked into it."

In the first days, every person pitched in, managers and supervisors alike. The higher-ups drove trucks and tossed garbage and waste piles into the backs of trucks. "Nobody worked harder than Rooney," Blew said. "I still remember seeing him out on the street loading piles of garbage. Everybody was out trying to get past that initial period." Every person had a job to do. Construction personnel supervised cleanup operations; accountants helped sweep streets. Rooney hired a local contractor to supplement the team until more equipment was ready and manpower recruited.

The service also included eliminating wild dogs, which were dispatched using blowguns to fire tranquilizing darts. Guns were not permitted.

Melk and Koenig moved over to operations at the start-up. "Sheikh Abdullah, he was the mayor," Weinert said. "This was his baby, and John Melk worked mostly with him. He was holding us to the highest standards possible, which was good for everybody. When we started, we stumbled."

Indeed, Mayor Abdullah had high expectations for his city's cleanliness, Falkman said. "He would take an ordinary light bulb and break it and place a piece of the bulb's glass against the street curb and tell us, 'You're missing, you're not cleaning, you're going by, you're not getting the glass.'" It was his way to get us to thoroughly clean the city. And you could almost eat from the streets in those days. It got very clean."

On another occasion, the mayor was driving Buntrock around the city. "He'd find a cigarette butt or cigar on the street. He'd complain. That's how they'd keep us on our toes," Buntrock said.

GROWING PAINS

Cleaning operations soon normalized. However, a contract dispute arose within a year that nearly ended the work. The Watson Saudi Arabia engineering firm had inserted a map of Riyadh marked with a dark line encircling the city that stated in bold letters "contract boundary" in the tender specs.

The mayor ordered that waste be collected in the areas outside the boundary, where the city was growing rapidly. The partnership refused, claiming it was beyond the agreed-upon city boundary on the map. The mayor argued that the partnership was nonetheless responsible for collecting all the growth in waste in the city over the five-year agreement. There were references to that in the contract. The dispute ended up going to the Saudi Board of Grievances, a court outside of Sharia law, where, ultimately, because of the boundary map, a settlement was reached with the partnership agreeing to collect in a few areas outside the boundary.

Meanwhile, in the Camp, cultural issues arose. Employees from one region complained about the cuisine prepared by chefs from another region. On one occasion, some took to the Camp's streets to complain. They found Rooney and demanded he taste the food himself. "'Bob, we're heading for the mess. We're going to eat some of their food,'" Keleher recalled Rooney saying.

At the time, the Camp was not yet completed. A prefab metal building at the site's center served as a temporary mess hall. The workers had seen photos in a recruitment brochure that depicted a completed facility with pools and landscaping, and it was not what they were experiencing. The workers decided it was not going to work even as Rooney grabbed a megaphone to calm them down. That night, Saudi officials intervened, and calm was restored.

The day the swimming pools opened, tragedy was narrowly avoided. Blomberg arranged to snap Polaroid photos of employees swimming to send home to their families in India. Popovich and Wayne Thurman of Bechtel, who later joined Waste Management, jumped in along with three Indian construction office clerks. One of the clerks had never been in a pool and was suddenly thrashing about, believing he was in over his head. Realizing he was not acting, bystanders jumped to his rescue. He was saved, and immediately afterward, hundreds more jumped into the four-foot-deep pool in celebration.

Employees sometimes used the pools for other purposes, too. The day of the opening, night managers walking past a pool saw a few workers washing their clothes and brushing their teeth in it. Some employees had come from villages where it was customary to bathe in a river with a bucket of water. Some workers avoided getting water on their heads. Waste Management's team removed some of the Camp's shower heads and pipes so they could splash water on their bodies by hand.

Collecting waste on Riyadh's unnamed streets also presented problems, particularly in routing vehicles. The team became creative, coining nicknames for key streets. There was Chicken Street, named for its multiple rotisseries, and Television Street, named for its numerous TV stores. Another street was called Caudle Causeway, named for the project's operations leader.

The team introduced an anti-littering program to Riyadh's residents to help with the cleaning effort. It relied heavily on educational materials from the US nonprofit's Keep America Beautiful program. Signs were posted throughout the city depicting an Arab man in a white thobe and ghutra placing litter in a container. A downtown billboard displayed a fluttered message to "Keep Riyadh Clean." Schools taught children anti-littering messages that could be carried home to their family members.

Dale Hoekstra was twenty-one and training in landfill operations at the CID Landfill in Chicago when his site manager, Dick Molenhouse, asked if he wanted to make some extra money in Saudi Arabia. Hoekstra, who had had some training in running equipment operations, arrived in Riyadh expecting to operate heavy machinery. Like many who arrived, he quickly learned to be flexible. The operations center assigned him to repair the street signs depicting Arabians dropping waste in garbage cans. Some needed removal, some repair. A "big red-headed gentleman" gave him his orders, he said, telling him to "drive through the city and do what you can to make these signs better." Hoekstra had just met Rooney. Hoekstra wasn't in Saudi Arabia long, but he would spend the next forty-four years with Waste Management developing and leading landfill operations before retiring in 2021.

RAINING "RAISINS" IN RIYADH

The team encountered a few environmental and technical issues in the process of cleaning, too.

For one, the working conditions and climate in the desert were difficult for those not used to the climate. While winter was mild, with moderate day temperatures and cold nights, summer averages between June and September when there was no rain could rise to 120 degrees.

Riyadh's driving regulations were also not enforced, and its street-lights were often not working. Uncoordinated infrastructure work also didn't help. A road would be newly paved, only to have a telephone company dig a trench for cable, followed by the electric utility that would do the same. New streets were filled with saw cuts and sometimes open trenches, making driving at night particularly hazardous.

Few locations had telephones, which made it difficult to arrange meetings. Connecting with family back in the States was harder. Waste Management's employees had to book telephone calls through the Saudi operator on a Friday, their day off. The wait could take hours. The operator would ring the caller back and connect them with family. The call would last only as long as the operator allowed, until breaking in and disconnecting it.

Insects were plentiful. Waste Management's landfill was next to an area local contractors used to illegally dump animal waste. Workers were forced to wear masks to shield them from inhaling and ingesting flies. Popovich arranged to pump thousands of gallons of fuel into the site to incinerate the remains. It burned for almost a month, Keleher said.

When the municipality hired a plane to dispense insecticide, flies dropped from the sky "like it was raining raisins," he said. Keleher recalled a trip to the landfill before construction where he saw a brown bulldozer that had just been delivered. Popovich tossed a stone at the dozer's blade, causing thousands of flies to flee and reveal its true greenish-yellow color.

But even in the midst of these factors, the long-shot bid in Riyadh had provided the Waste Management team with valuable experience and, more importantly, confidence. The team had delivered on the contract and done the work. Riyadh was the clean city that the Saudis had sought.

After five years, the team would leave to take on new challenges. Jerry Rhodes would assist in other international surveys and go on to manage the company's chemical waste operations in Coalinga, California.

Mike Curry, a mechanical engineer, and his wife would move to the United States and also join the chemical waste management business. Curtiss Collins, the J.A. Jones construction manager, would join Waste Management, and a few years later, supervise the building of another Saudi facility. Martin Weller, a joint venture manager from the United Kingdom, would work on other international projects. Mick Witnell, a UK-hired sweeping supervisor, became the company's international sweeping expert. Jim Koenig, an operational leader, would go on to lead the development of the company's Superfund remediation group and eventually rise to become the company's chief financial officer. Weinert would become president of Waste Management International. Falkman would become its general counsel and succeed Weinert later as its president.

In the meantime, the next international adventure awaited: South America.

TEAM MEMBERS LISTED IN THIS CHAPTER

Dick Beck	Ron Markham
Ian Bird	Larry Norton
Dave Blomberg	Joe Parziale
Dean Buntrock	Bob Paul
Dr. Emilio Cardenas	Margarita Porcel
Jerry Caudle	Ron Reed
Dave Coleman	Jerry Rhodes
Bill Debes	Phil Rooney
Art Dudzinski	Roland Schmidt
Greg Fairbanks	Ron Shufflebotham
Ed Falkman	Eduardo Terreni
Don Flynn	Bill Thon
Diego Galdon	Bruce Tobecksen
Jerry Girsch	Bob Van Tholen
Alex Gonzales	Jim Waters
Derek Irlam	Fred Weinert
Bob Keleher	Martin Weller
Peter Loganzo	Mick Whitnell

12

ADVENTURE IN BUENOS AIRES

In May 1979, Waste Management's international team sprinted out of the Saudi desert filled with confidence. They had taken on a huge challenge in even pursuing the bid to clean Riyadh. Now, they would shoot even higher.

Fred Weinert, who had directed construction in Riyadh, was working hard to collect the final construction payment from the Saudis when he was called back to the waste industry's convention in Kansas City. He thought it would be a quick trip. He had work to do and zero interest in going to the convention. "I literally left there with nothing other than a short-sleeve shirt and a passport because I [thought I] was coming right back," he said.

Waste Management had other plans for him. "We want you to look at a project in Buenos Aires," Buntrock told him. Buntrock had learned about a potential project there from a chance meeting at a cocktail party at the American Embassy in Jeddah. His good friend George Johnson knew the US ambassador and had come along. Buntrock found himself in conversation with the Argentine ambassador, who described the problems Buenos Aires faced in managing its waste and invited him to send a company team down to take a look. He did and asked Weinert to lead the quest.

As Weinert soon learned on his first visit to Buenos Aires, the opportunity was massive. The Argentines wanted to privatize the city's garbage collection and street sweeping—the first privatization in the city's history. The potential contract was worth double what was won in Riyadh.

A few years earlier, Waste Management had sought and failed to win a transfer and disposal contract in Buenos Aires, which would have been easier work for the company. Still, the team was optimistic. "Put together the team, let's do it," he was told. Weinert thought it would be big.

A BREATH OF FRESH AIR IN BUENOS AIRES

Buenos Aires officials were thinking big, too. The project was key to their development of the Coordinación Ecológica Área Metropolitana Sociedad del Estado (CEAMSE), or "The Green Belt." CEAMSE was owned equally by the province and the City of Buenos Aires and created to modernize sanitation services for the city's millions of residents, providing a better system of waste collection and transfer and disposal services for the waste generated daily. "The name basically symbolized the creation of a forested area around the city that would capture carbon and emit oxygen in the air to give the city cleaner air to breathe," said Falkman.

The Buenos Aires mayor, Osvaldo Cacciatore, a former Argentine Air Force officer, and CEAMSE initiated the project to clean the city and called for an international tender. Buenos Aires was dirty and its air grimy from the pollution of tens of thousands of residential waste incinerators. There had also been a surge in auto traffic and the city's population as migrants journeyed from the country's less prosperous areas.

The project came during a difficult political time in Argentina. Tensions remained taut, and the country's blood pressure was high. In 1976, military chiefs seized power over the country and city after a coup ousted former president Isabel Martínez de Perón. Leftist guerilla groups called *Montoneros* fomented violence and protests. The country was experiencing skyrocketing inflation, and rising price tags on consumer goods were pasted atop the previous day's prices daily. The times were uneasy and dangerous in Buenos Aires.

But Cacciatore was committed to modernizing the city's infrastructure and public works systems. He had just the person to get it done.

"There was one man behind it: Dr. Guillermo Lauer," Weinert said. "He was an admirer of public works in the United States." Cacciatore appointed Dr. Lauer, a civil engineer, to lead the city's transformation. Lauer thought big, too. There would be new highways, the demolition of slums and construction of new housing, new parks, new public schools, and a new environmentally sound waste management program. The Green Belt they foresaw would literally be a breath of fresh air.

AN ENTREPRENEURIAL PARTNER

For Waste Management, the first step was to find a partner.

After a search, the executive team reduced the list to three candidates: a well-established financial group, an international logistics group, and an industrial/construction group. Franco Macri, a contractor and prominent entrepreneur, was a match. Macri, a Roman-born industrialist, had a history of aligning himself with leading international groups. He was successful and experienced. He had started out building homes and completing small public works jobs. With the country's economic recovery in the mid-1960s, he became a major contractor, constructing a factory for Fiat, the Italian automaker, and infrastructure projects such as power plants, bridges, highways, and towers. He supported the arts. He introduced the first cellular phone in Argentina, acquired a series of important electronic businesses, and established the construction group Impresit-Sideco. The latter would be Waste Management's partner in creating Mantenga Limpia a Buenos Aires (MANLIBA), or Help Keep Buenos Aires Clean, a joint venture established to pursue the Buenos Aires contract.

Macri was also chairman of the local Banco de Italia. "He branched out in other businesses with the objective to always be a partner with an industry leader in whatever business he was pursuing," said Weinert. "He wanted to be with Waste Management, the No. 1 environmental company in the world, as far as he was concerned."

Jerry Girsch, another numbers man from Arthur Andersen, had been enlisted to help find a partner. He later negotiated the partnership. His presence was important, and he made ten or so trips there.

On one of his first trips, he and a company delegation arrived at a classic building in the business district. A dated, gated elevator transported visitors to their floors. A small-statured woman was seated at the controls. After accepting the carriage's occupants, she would slide its grated gate closed and then move a hand lever left or right to go up or down.

Girsch was a man of substantial stature and girth. The men joined others on board. Girsch joined last. The door closed, and the woman tugged the lever. Nothing. The carriage was frozen. "Finally, she says something," Girsch recalled. "Gringo grande!"

She opened the gate, and Girsch stepped off. The elevator rose and then returned. "There I am scratching my head," Girsch recalled. "Nobody else showed up. I got on the elevator. She has a big smile."

The project had to meet certain accounting rules that Girsch knew, standards that Flynn demanded. Girsch did well. Waste Management achieved 60 percent of the equity and a technical service fee of 5 percent of the revenue. The company also warranted that the waste would be collected and the streets swept, a no-brainer for the work it sought. Macri saw to it that his Banco de Italia provided debt capital. Waste Management and Macri could not have reached an agreement that was more mutually beneficial, the team believed.

Falkman was the team's general counsel. He ensured the proposal complied with everything the tender required. He also was responsible for all the contracts related to the project, including the Macri partnership and the agreement with CEAMSE.

But before Waste Management could bid, its team needed to know the size of the work and how to manage it. Weinert quickly put together a budget. After Riyadh, he knew what he needed. No one batted an eye at the survey's $250,000 price tag to determine the scope and price of the work. It was a bet on a "maybe," Weinert believed. Weinert assembled his survey team. If they won it, an operations team would follow.

REMAPPING BUENOS AIRES

A flight of senior executives, including Buntrock, Rooney, and Flynn, accompanied by a group of a dozen or so others, landed at the Buenos Aires airport after an overnight, 4,400-mile flight from Miami. It was the start of their mission to provide the best estimates of the city's daily garbage volumes and figure out how best to collect it from the city's communities. They did not yet know the city's landscape. Unsurprisingly, many of the team members were armed with Saudi Arabian experience and know-how in estimating waste volumes.

The team moved into the city's downtown Sheraton Hotel and set up shop in the ballroom. They pushed rectangular tables together to form one big table and spread a huge map of the city over it to begin their work. They worked nonstop there for three to four weeks.

Two team members were map writers toiling atop the table making corrections with pens and rulers. They needed to know how the streets were configured to inform collections and sweeping. And they needed to understand the city and the physical nature of the streets, their surfaces and widths, and whether there were curbs or sidewalks. In street sweeping, curbs are needed so litter can be pushed down the street and then scooped into a truck.

The streets of Buenos Aires reflected the city's European style. They were circular, narrow, unpaved and paved, some curbed, some with sidewalks. Many were cobblestone. Cars were parked everywhere, including sidewalks, which could stymie collection routes.

"You couldn't send a sweeper down a dirt street," said Bill Debes, a survey leader. "The other problem was how narrow they [the streets] were. You couldn't have the big packer trucks going down the narrow streets because you'd scrape the cars on both sides of the streets. We had to have an understanding of how the streets were configured, where the intersections were, and what the turns could be, and what we could do for cleaning the streets and all that kind of stuff."

In understanding the streets and potential routes, the map writers began remapping the city. The team used this information to put together a plan based on the most accurate data they could gather. It would determine routing as well as the cost of equipment and people to perform the task. Garbage was collected six nights a week. Saturday evening was the only time off.

Rooney, Debes, Blomberg, and Keleher formed a committee, joining with and directing the twenty or so others, including Jim Waters from Florida, Roland Schmidt, Peter Loganzo, Dick Beck, Ron Markham, and Derek Irlam, an Englishman. He had been involved in one of Britain's first companies pursuing industry consolidation and was retained as an international advisor.

One of the most important survey team members was Alex Gonzales, a Cuban-born refugee who escaped the island with his mother in 1956 before the Castro revolution. Gonzales's father had been in the Cuban revolutionary forces, and his family had been subjected to harassment by the government of dictator Fulgencio Batista y Zaldívar. As a child, Gonzales visited his father, a political prisoner, in a Cuban jail.

Wayne Huizenga identified Gonzales at a Florida Resource Recovery Council meeting in Cocoa Beach. They lunched over sandwiches and potato chips, and Huizenga offered him a job on the spot. Gonzales turned it down, but before long, found himself in the company's Pompano Beach office with Huizenga and manager Earl Eberlin. They were too persuasive, and he couldn't resist their offer.

Gonzales initially worked as the Buenos Aires team's liaison to the government. It was his job not only to translate but also to communicate and smooth over any problems. He was good at it. "He was very instrumental for me in the bid process because he is a great communicator and personality, and he knew the business inside and out," Weinert said.

Gonzales would leave part of himself in Argentina after he was stricken with appendicitis and had to have surgery. Following the procedure, he was flown first class back to Miami and hospitalized again there to recover.

SURVEYING FOR GARBAGE

The survey teams had to learn how many tons of waste needed to be collected every day to decide how much equipment was needed. They found the garbage reflected the society: Most residents lived in small apartments with small refrigerators in high-rises. They bought only fresh food required for a short time and discarded food scraps, making their garbage heavier, full of "wet" waste. They had little packaging as most cardboard was scavenged.

"It was very common for a family to buy a dinner with the vegetables and the salads and that kind of stuff that day, throw it out that night, and buy another for the next day," Debes said. "The garbage was very green." Hog farmers surrounding the city even had permits to pick up the garbage from restaurants in the central business district and use it to feed their hogs.

Many of the buildings' front doors faced the streets. The more modern residences had eight-to-eighteen-story high-rises with upwards of 120 apartments. There could be one 100 doorbells in them, too many and too many blocks to count. For the survey team, the small apartments were a clue for waste generation that the residents would not contribute as high a volume of waste.

Residents would dispose of their waste in small bags placed into baskets attached to bicycle stands. The baskets were a few feet above the pavement to prevent rodents from getting at them. Dogs, however, often found their way into the bags. Sometimes residents would just bring their waste to an intersection and leave it there to pile up and be collected days later. Larger buildings had compactors installed that replaced incinerators. The compactors packed up to two hundred pounds of waste into plastic tubes that were pitched into trucks and disposed of at CEAMSE sanitary landfills. The fewer incinerators meant clearer skies.

Debes created a form to record the volume estimates. The existing collection system was not as efficient—it had been developed as a government make-work project designed to employ as many people as possible.

The survey teams paired up with and were aided by a group of about twenty medical students who joined the effort, serving as translators and often, city guides at night. "They were invaluable," Debes said. "Not only did they know the language; they knew the city. If we wanted to go to a certain area, they could get us there in the quickest way."

The team carried special passports, the result of the Argentines' efforts to bring order during a combustible time when nerves were raw and identities needed to be clear. They also carried letters identifying them as government contractors and referring any people who might stop them to the mayor's office. One crew member was detained, but the right papers and quick-thinking team members retrieved him. The papers were "a get-out-of-jail-free card," Debes said.

During the midpoint in the survey process, the team came across a critical document while in conversation with local authorities. The document contained the municipality's 1970 census data and revealed how many people lived in a square block—giving the team valuable data they could then use to more accurately estimate refuse weights and volumes on a block-by-block basis. With the document in possession, Keleher played a key role in summarizing the information. "It became the basis for the survey," Weinert said. "It allowed the team to estimate the garbage volumes and plan the collection routing and productivity measures."

THE BIDDING BEGINS

On August 29, 1979, the tender went public. A dozen companies responded: Two of them were longtime US rivals—BFI and SCA Services, and the others were less experienced operators. (Waste Management's teams would often encounter their American counterparts and interpreters at local restaurants during the bid process. "We'd be staring at each other," a Waste Management team member later said.)

The bids were opened in two stages. The first eliminated unqualified companies. The second concerned only the companies deemed to have qualified operating plans. At this stage, only price mattered. Waste Management's bid was several cents per ton lower than the next company's.

A month later, the Waste Management–Macri team won the CEAMSE tender and was awarded residential and industrial collection in four zones with more than two million people, as well as street sweeping and flushing and catch basin cleaning. Residential collection and street cleaning would occur six nights a week, and commercial collection and the removal of corner waste piles were done during the day. The contract was valued at $400 million—$40 million a year over ten years.

Dr. Emilio Cardenas was a critical figure to help the company negotiate the contract. A consummate professional, he was a well-respected local lawyer people listened to, a man with influence. A professor in Buenos Aires (and at the University of Illinois law school), he knew everyone wherever he went. On one occasion he was late for a negotiating session with CEAMSE. "He knows there's a list of twelve items we wanted to get into the contract," Weinert said, recalling that the Americans had failed to make any progress. "He walks into the room, everyone knows him. We obtained ten of the twelve items. He is listened to. He built a reputation." Cardenas would later become Argentina's ambassador to the United Nations representing the country on the international body's Security Council.

Within weeks the contract was signed. Waste Management had until March to start, just twenty-one weeks.

MOBILIZING A TEAM AND GARBAGE COLLECTION

Now the joint venture MANLIBA had to deliver on its promises and execute. Its leaders laid out mobilization plans and purchased equipment. With start-up ahead, more team members arrived, many with valued Saudi experience.

Starting in late 1979, Weinert began leading the mobilization, with Falkman as his deputy. Jerry Caudle was pulled out of Riyadh (replaced by Art Dudzinski) to oversee operations and the project's largest truck depot in preparing for the launch. He was aided by Gonzales, who was moved over from the survey team, and Jerry Rhodes, who helped with the facilities' setup. Joe Parziale, an experienced logistics hand, proved pivotal in shipping equipment from the US to Argentina at a favorable customs rate.

Bruce Tobecksen, a new recruit from Arthur Andersen who would later become a senior financial executive, took on the role of finance director for the mobilization, responsible for setting up MANLIBA's accounting department and controls, and installing its financial systems and reporting. His task, a difficult one under any circumstance, was made even more onerous by Argentina's hyperinflation rate that at one point hit 1,000 percent annually and resulted in almost daily currency devaluations.

Don Flynn would always ask Weinert, "Do you need anything?" Weinert needed a smart financial guy willing to be on-site full time. Flynn "reaches into his Arthur Andersen barrel," Weinert recalled. He identified Greg Fairbanks, then still at Arthur Andersen in Florida but who would become a significant addition to the MANLIBA accounting and financial team after start-up. He oversaw and trained a talented team hired locally and eventually took over for Tobecksen when he returned to the United States.

Bob Paul flew down to set up warehousing and receipt of equipment from overseas. Bill Thon aided equipment purchasing. Ron Shufflebotham, the company's steady hand in public relations and promotion, led the prepping of the new fleet with MANLIBA decals. Bob Van Tholen from corporate and Larry Norton from Alabama assisted with trucking.

Several employees also came from Britain: Martin Weller set up vehicle maintenance with Jerry Rhodes, Mick Whitnell oversaw street sweeping, and Ron Reed developed a procedures manual. Ian Bird was recruited to be in-house counsel during the start-up period and went on to have other international assignments.

Diego Galdon was another key addition. An Argentine, he came from Ford and was critical in establishing an effective labor relations program. Dave Coleman assisted with union negotiations. Eduardo Terreni, an Argentine also from Ford, worked under Caudle managing another truck depot. He was like an army sergeant. On the day of the launch, his fleet of trucks was lined up and ready. They were out of the gate in two minutes. Everything went according to plan—at first.

A CHALLENGING START

On March 2, 1980, only five months after the contract's signing, MANLIBA was on the streets of Buenos Aires with a sparkling new fleet of 180 white Truxmore and Coby side-loader refuse compactors imported from the US and mounted on Fiat trucks made in Argentina. They were supported by a fleet of Johnson vacuum street sweepers and Vac All catch basin cleaners and employees whose number would eventually reach 1,500. The MANLIBA workers, unlike their predecessors, wore crisp uniforms and protective gloves.

A few weeks after launch, however, there was trouble. MANLIBA had disrupted the old Buenos Aires refuse collection system. The cost of cleaning the city had ballooned during the Perón years. Gone were the legion of 8,000 to 9,000 people who had been cleaning the city, a multitude above the current number. Gone, too, were the small trucking firms that had been earning millions of dollars removing the residents' corner waste piles. Disruption meant eruption.

"There was a lot of resistance," Falkman said. "There were a lot of people who lost out. For the leftist *Montoneros*, the armed resistance group, this was a direct affront to them privatizing what they considered a public service. We didn't inherit the city workers because this wasn't a city contract. We were a new private company, and we started from scratch."

On a hot Friday night, three weeks after the launch, rumors surfaced of an attack on MANLIBA crews. The collection crews were reluctant to go to work but were persuaded to start their regular 9 p.m. to 5 a.m. routes. Two hours later, at 11 p.m., calls came in that trucks had been shot at, tires blown out, and windshields smashed. A few trucks were hijacked. Around midnight, MANLIBA suspended operations and ordered the trucks and crews to return to their depots. They would be off duty on the following night, the only time when there were no collections.

Weinert alerted Macri, who came to the main garage and learned about the situation. He asked what the company needed. How could they relaunch collections on Sunday night? The union also wanted protection for the crews and wouldn't work without additional security.

Weinert and Macri developed a plan that called for a third garage that they could dispatch trucks from.

Macri (whose son became Argentina's president decades later) asked Weinert to be in the office the next morning. He would call him with directions. The next day, he told Weinert to meet him at a government building with a large rotunda to discuss the issue with authorities. Initially, the officials were accusatory: "Did you cause this to happen, you may have created this?" Weinert recalled them asking. Once their questions were satisfied and requests were made, the authorities agreed to MANLIBA's security requests: They embedded an armed policeman with a two-way radio in every truck. They assigned police cars to shadow them. The police cars also surrounded each garage, on guard for trouble. The protection continued for three weeks while the police investigated the attacks, identified who was behind them, and contained the problem. In a matter of weeks, authorities arrested a ringleader.

INVESTING IN THE COMMUNITY

MANLIBA understood the value of a strong image, and the partnership invested in programs to strengthen its ties to the community. To help ensure the program's success, buy-in, and image, MANLIBA launched a public education and community relations campaign led by Margarita Porcel, an Argentine who had previously worked public relations in Kuwait.

Porcel had both the talent and budget for public relations and advertising. She hired a team of women representing MANLIBA who were in the city's communities every day handing out flyers and information about keeping Buenos Aires clean. They worked with neighborhood associations and raised money in the community. They worked internally, too, to organize an end-of-the-year dinner for all the 1,500 to 2,000 employees at a soccer stadium.

The community relations program encouraged residents to play a role in supporting cleanliness, health, and the environment, Weinert said. They targeted neighborhood associations, schools, festivals, social groups, influencers, and the media to help spread this message. With these efforts, MANLIBA earned a lasting positive image. Additionally, the MANLIBA community relations program earned special recognition by the NSWMA at its annual meeting, where Porcel made a presentation about the program. "A top BFI executive approached me and issued a friendly warning that he was going to try to steal Porcel away," Weinert remembered.

As the contract was underway, there was near tragedy. The team was working out of leased office space on the twenty-fifth floor of a Macri building. The building's maintenance crew was working on an elevator when a fire ignited in it and smoke and flames shot up the shaft. It created an inferno. Although the fire occurred over a holiday weekend, Caudle and a half-dozen others were at work. Their only escape route was a stairway that wrapped around the flaming elevator shaft. With the stairway ablaze, they tried the windows. The windows would not open. The team decided to put coverings around the doors to block the smoke and then waited for rescue. Firemen finally reached the team and escorted them to safety. Everyone was safe, but the team no longer had offices. They moved briefly back to the Sheraton Hotel and then quickly purchased two floors in a building not far from the site of the fire. When they moved in, the building's construction had not yet been completed, and there was no heat during the Argentine winter. The employees continued their work wearing their winter coats.

MANLIBA's benefit to Buenos Aires and its residents grew more apparent: The company saved the city $50 million a year and left the city's streets pristine. MANLIBA-trained employees served all 600,000 households, apartments, and commercial shops. Some 500,000 curb kilometers of streets were swept annually across the city and 32,000 catch basins cleaned. The MANLIBA team delivered. When the contract's term ended in 1990, CEAMSE exercised options to extend the contract for another six years.

In the recent past, Galdon and Weinert met one of MANLIBA's main union delegates from the early years. "Today he is the most powerful union leader in Argentina," Weinert said. "He said in so many words that MANLIBA changed the waste collector's job and profile as never before. We respected and valued the workers and treated them with dignity, setting a new and positive standard for the refuse worker in Argentina."

Waste Management's international journey was only beginning. They would next provide waste collection, street sweeping, and landfill services to the one million residents in Cordoba, Argentina. Down the line, they would lead smaller projects in Bariloche and Bahia Blanca, Argentina, and at a landfill outside Buenos Aires.

Before long, the South American team's MASURCA operations won an eight-year contract to operate in a part of Caracas, Venezuela. They served 350,000 residents, the Simon Bolivar International Airport, and the La Guaira seaport.

With success sealed in South America, the team turned its attention back to the Middle East. Saudi Arabia had another project.

Chris Disbrow (left), a senior district manager, and drivers Glenda Schaller (center) and Bob Marcione (right), who has since retired, at Waste Management's Cicero, Illinois site. *Photo credit: Courtesy of Chris Disbrow*

Waste Management's senior executive team in the early 1970s included (standing from left to right) Don Flynn, Hal Gershowitz, Larry Beck, and Phil Rooney, and (seated) Dean Buntrock and Wayne Huizenga. *Photo credit: Buntrock family archive*

Wayne Huizenga's Southern Sanitation's waste collection trucks in Ft. Lauderdale, Florida, facetiously advertised "free snow removal" and "We cater weddings." *Photo credit: Buntrock family archive*

Waste Management's early financial leaders included (left to right) Controller Jerry Girsch, Vice President of Finance Don Flynn, and Treasurer Bob Paul. *Photo credit: Buntrock family archive*

Dean Buntrock (left) and Wayne Huizenga (right) visit the Riyadh, Saudi Arabia, camel market during a Waste Management mobilization planning trip in the mid-1970s. *Photo credit: Buntrock family archive*

Candidates in Mumbai, India, line up for interviews for Waste Management's city-cleaning contract in Riyadh, Saudi Arabia. "We had literally thousands of people coming in over a short period of time," said Human Resources Manager Jim McGrath. *Photo credit: Courtesy of Jim McGrath*

The 1977 Riyadh contract called for the construction of a self-contained Camp to house and support more than 2,000 employees, manage site operations, maintain and dispatch a 1,000-vehicle fleet, and provide disposal capability. The Camp, which had to be built in less than a year, also had mess halls, a mosque, laundry, commissary, and a clinic. *Photo credit: Buntrock family archive*

Waste Management's Saudi Arabian mobilization teams launched a public education program that urged residents to "help make Riyadh clean." *Photo credit: Buntrock family archive*

Waste Management's $400-million, ten-year joint venture contract in Buenos Aires, Argentina, served 600,000 households and provided waste collection, street-sweeping and flushing, and catch basin cleaning six days a week. The service, which began in 1979, operated under the name MANLIBA, which stood for Mantenga Limpia a Buenos Aires, or Help Keep Buenos Aires Clean. *Photo credit: Buntrock family archive*

The contract in Buenos Aires called for MANLIBA teams to clean the city's catch basins. A team that eventually grew to 1,500 workers used a new fleet of trucks to collect trash, sweep streets, and vacuum catch basins. *Photo credit: Buntrock family archive*

Waste Management's executive team, who won a 1981 contract by royal decree to provide city cleaning services to Jeddah, Saudi Arabia, included (from left to right) Bill Reichert, Mike Rogan, Joe Jack, John Melk, and Ed Falkman. *Photo credit: Buntrock family archive*

The Waste Management camp in Jeddah, Saudi Arabia, included seventy-five acres and 320,000 square feet of buildings. The contract served more than a million residents and included collecting and disposing of waste, sweeping 40,000 curb miles of streets, collecting litter, cleaning public buildings and monuments, removing abandoned cars, and handling pest and animal control. *Photo credit: Buntrock family archive*

Dean Buntrock (right) with Jeddah's Mayor Dr. Mohamed Said Farsi (left) cut ribbon to launch the five-year renewal of Waste Management's contract in Jeddah, Saudi Arabia, known as the Jeddah Cleansing Project. Fred Weinert, the president of Waste Management International, stands in between the two. *Photo credit: Buntrock family archive*

Waste Management in 1984 gifted the city of Jeddah with a sculpture called Peace by Mexican artist Leonardo Nierman that was prominently placed in the city's Central Corniche along the Red Sea. *Photo credit: Buntrock family archive*

Waste Management arrived in Brisbane, Australia, in 1983, after winning a contract to provide waste collection services there. It won another seven-year contract in 1990, employing a new forty-vehicle fleet to service 137,000 residents within the city. *Photo credit: Buntrock family archive*

Waste Management's Pacific Waste Management subsidiary expanded across Australia, acquiring businesses in Sydney, Melbourne, Canberra, and Adelaide, becoming one of the largest waste operators in the country. *Photo credit: Buntrock family archive*

Waste Management employs environmental systems across its network of landfills, including engineered liners and leachate collection to protect groundwater, daily cover to control litter and pests, groundwater monitoring, and the recovery of gas from decomposing waste to generate energy. *Photo credit: Buntrock family archive*

Waste Management landfills are designed to provide for recreational end uses, including golf courses, parks, sports facilities, and wildlife habitats. *Photo credit: Buntrock family archive*

Waste Management's executives saw an opportunity to manage hazardous wastes after the U.S. Congress enacted the Resource Conservation and Recovery Act in 1976 and the Comprehensive Environmental Response, Compensation and Liability Act, known as Superfund, in 1980, which established new regulations on managing solid and hazardous waste. The company's Chemical Waste Management subsidiary became the nation's largest manager of hazardous wastes and most prominent remediator of contaminated waste sites. *Photo credit: Buntrock family archive*

Jodie Bernstein became Chemical Waste Management's general counsel in 1983 after compliance questions arose. She later developed the company's business ethics program. *Photo credit: Buntrock family archive*

Chemical Waste Management developed a national network of facilities to dispose of and treat chemical hazardous wastes. Its Port Arthur, Texas, site is shown here. *Photo credit: Buntrock family archive*

Managing the range of hazardous wastes was a more complex business than handling the collection and disposal of garbage. Chemical Waste Management built facilities that could test and track industrial wastes to ensure their proper treatment and disposal. *Photo credit: Buntrock family archive*

The *Vulcanus I* was the first of two ships that Waste Management hoped to operate in U.S. waters to incinerate hazardous wastes at sea. Despite successful burns of PCBs, DDT, and Agent Orange, the ships never sailed in commercial service in the U.S. *Photo credit: Buntrock family archive*

In 1992, Chemical Waste Management acquired Chem-Nuclear Systems, a specialized business that provided low-level radioactive wastes services. Chem-Nuclear's customers included more than one hundred ninety nuclear facilities, government installations, hospitals, and commercial laboratories. *Photo credit: Buntrock family archive*

Jane Witheridge, who was one of the first engineers Peter Vardy recruited to Waste Management in 1976, helped to pioneer the company's environmental audit program. She later led the operation of the company's disposal sites in Pennsylvania and directed its corporate Recycle America recycling programs. *Photo credit: Buntrock family archive*

Waste Management made recycling a priority in the mid-1980s. Its Empire Waste Management subsidiary in California pioneered residential curbside recycling. The volumes of recyclables—paper, metals, plastics, and glass—required increasingly sophisticated material recovering facilities called MRFs. *Photo credit: Buntrock family archive*

Recycle America material recovery facilities sorted collected recyclables and separated them into commodities for end markets. Baling machines prepared the materials for shipment to end users. *Photo credit: Buntrock family archive*

Waste Management's Wheelabrator Technologies Inc. subsidiary burned trash to generate energy. Wheelabrator's fourteen facilities generated 12 million kilowatt hours of electricity a day, the equivalent of 17,000 barrels of oil. *Photo credit: Buntrock family archive*

Waste Management expanded beyond waste services to include clean air and clean water technologies and environmental engineering. Its EOS water technologies subsidiary provided water and wastewater treatment capability. *Photo credit: Buntrock family archive*

Waste Management's executive team in the early 1990s included (left to right) Steve Bergerson, general counsel; Don Flynn, chief financial officer; Phil Rooney, chief operating officer; Dean Buntrock, chairman and chief executive officer; Jerry Dempsey, chief executive of Chemical Waste Management; Hal Gershowitz, senior vice president; and Jim Koenig, vice president of finance. *Photo credit: Buntrock family archive*

The families of Waste Management's founders celebrated the fiftieth anniversary of the company's initial public offering in June 2021. Attending the reception were (from left to right) P.J. Huizenga; Waste Management President and CEO Jim Fish; Dean Buntrock; Executive Vice President and Chief Financial Officer Devina Rankin; Senior Vice President and Chief Customer Officer Mike Watson; and retired Executive Vice President and Chief Operating Officer Jim Trevathan. *Photo credit: P.J. Huizenga*

TEAM MEMBERS LISTED IN THIS CHAPTER

Dean Buntrock	John Melk
Dave Coleman	Jim Nothnagel
Curtis Collins	Joe Parziale
Mike Curry	Dean Ramsey
Ed Falkman	Bill Reichert
Bud Ingalls	Mike Rogan
Joe Jack	Phil Rooney
Alton Jamieson	Tony Shahbarat
Rod Jarvis	John Slocum
Rich Jeffrey	Wayne Thurman
Tim Jessup	Bob Van Tholen
Bob Keleher	Dave Walker
Dennis Kiraly	Jim Waters
James Mears	Fred Weinert

13

BACK TO ARABIA AND ON TO
AUSTRALASIA

As the 1980s arrived, another international challenge was on the horizon. Saudi Arabia's royal family wanted to clean Jeddah, its growing port city on the Red Sea. They had been committed to investing in and modernizing the country, transforming Jeddah into a clean, modern city. It would quickly become one of the most important cities in the Kingdom, growing from about 380,000 in the early 1970s to nearly a million residents in the early 1980s.

The mayor of Jeddah, Dr. Mohamed Said Farsi, was an architect and art lover. He was also a visionary. The city's buildings were made of white coral stone block; they reflected the turquoise blue sea. Farsi knew what he wanted: a clean, modern city.

Waste Management, already proven in Riyadh and Buenos Aires, turned its attention back to the Middle East.

It would be another large opportunity for the company, which now pursued its overseas expansion as Waste Management International with its Arabian Cleaning Enterprises (ACE) team, adding many millions to the company treasury. And it would be another challenge. But this time, its leaders came armed with experience. They knew what to do.

The ACE partnership gained the contract by royal decree in 1981 following a rigorous bidding process. The work would be much broader and much bigger than Riyadh. ACE would collect garbage and rubble from a million residents, sweep more than 40,000 curb miles of streets, clean more than 300,000 square feet of buildings and public monuments monthly, clean public toilets, remove thousands of abandoned cars, and handle pest control, including roving packs of dogs, the latter almost more than one could imagine. The team would also build a seventy-five-acre camp with maintenance facilities for as many as one thousand vehicles and provide housing, food service, recreational facilities, and amenities for thousands of Indian and Sri Lankan workers.

They had just twelve months to build the Camp and start cleaning. Pricing the work was complicated. Unlike Riyadh, there was no fixed price. The head of Jeddah's public works disliked fixed pricing. The way he saw it, a low, underpriced bid would result in the contractor cutting corners and providing unsatisfactory service to the detriment of the city. A high price meant the city would be charged too much. The public works official devised a pricing system that included as many as eleven categories with waste volumes estimated for the kind of housing— homes, apartments, and so forth—contained in each category or city area. The pricing system required aerial photography and mapping city blocks. It was complex and difficult. And it failed to reflect the number of residences in denser areas, resulting in disputes that necessitated agreement amendments and took long periods to resolve.

Melk, Falkman, Weinert and Rogan—the team who helped stand up the Buenos Aires project—worked on the contract. On the ground, a team began preparing for the work ahead.

THE JEDDAH TEAM

Bill Reichert and John Slocum arrived in November 1981 after the contract's award. Reichert, the *el jefe*, or leader, ran the site's operations. Slocum was the senior finance director. He watched the money. He often chased after it. Dennis Kiraly came on to help with the accounting. Jim Waters, a company veteran from Florida, assisted in operations and later became the Saudi country manager.

Slocum hired Tony Shahbarat, a Jordanian who had gained some sports fame as a star soccer-style place kicker for the University of North Dakota, where he earned an accounting degree. He spoke Arabic. Shahbarat took on a number of jobs, perhaps most important, liaising with city officials. He also became involved with hiring the site's workforce, helping James Mears, who had recruited Riyadh's workforce, attract candidates in Bombay. At one point, Shahbarat and Mears learned the employment company they had hired had been charging recruits for the jobs. Shahbarat went to Bombay to set things right. In the months following, he and Slocum worked together closely. Shahbarat also got along well with the workers, the footballer often playing soccer with them.

Slocum, Reichert, and their wives were once invited to meet the Indian Prime Minister, Indira Gandhi, who wanted to thank them for their efforts in protecting employees' rights. She was aware that some companies treated Indian workers poorly, Slocum said. "She was wonderful. Meeting her was the highlight of my life. She couldn't say enough about our company and how proud she was of what we did."

Joe Jack, another senior company executive from the States, arrived to lend a hand with the mobilization and construction. He was a troubleshooter, a helpful quality as projects like this always brought surprises. Tim Jessup, new to the company, handled truck maintenance. Rod Jarvis, in Oak Brook, was the primary purchaser.

About a half-dozen Americans brought their wives and children along. Separate housing was built for them. As women weren't allowed to drive in the country, a driver was provided for them and their families.

BUILDING A CAMP IN THE SAND

Bud Ingalls, a former Andersen accountant who had been focused on strategic planning in Oak Brook, managed construction of the seventy-five-acre camp and its 320,000 square feet of buildings. He had no prior construction experience.

For Ingalls, the biggest challenge was adapting to the environment.

The campsite was distant from the city's center, in the middle of the desert. "It was just sand," he said. "Nothing was there. No power, no roads. There were camels running around. Everything had to be built. We had to build roads to get to the facility. We had to work with the city on the design to make sure they were comfortable with it," he said. Ingalls reported to Rooney, who would fly over once a month to oversee the team's progress.

Bob Keleher, a more experienced construction hand, traveled from Oak Brook regularly to assist. The design and construction firm Bechtel was involved early on in the camp's design and construction, but Waste Management took over that work and brought in Donahue Engineering for the design. The construction team included experienced hands such as Mike Curry, Curtis Collins, Dean Ramsey, Jim Nothnagel, Dave Walker, and Wayne Thurman.

Unlike in Riyadh, the purchasing team could source much of what they needed in the Kingdom, Keleher said. "We didn't need to bring everything in from the outside."

And like in Riyadh, the camp—actually a small town—would have considerable infrastructure, including its own electricity generating plant, water and sewage treatment, a clinic and commissary, a warehouse, and a fuel depot. The buildings were made of finished white-faced concrete slabs and insulated. They would later build a landfill and a mosque.

WILD DOGS AND PILED-UP CARS: START-UP BEGINS

In 1981, the time for start-up began. The ACE team was not totally prepared. As in Riyadh, everyone pitched in on the streets and in the trucks.

Ingalls was given a truck and three helpers. "We went out at three-thirty or four in the morning and got back at eight or nine at night. It was a brutal period of time," he said. "We all had our trucks, would work all day, and would get back and die."

The waste was piled up on the streets and had to be shoveled into the trucks. The difficult period didn't last long for the managers, though. The start-up was phased, and in time, the team was better organized, and the cleaning crews were on the streets working according to plan.

Packs of wild, rabid dogs were a major problem in Jeddah. "We removed tens of thousands of dogs," Falkman recalled. "They roamed in packs and if you were out jogging in the morning and came across a pack of wild dogs, that wasn't a very comfortable experience, so they got removed."

Cars had also piled up over the years. The company picked them up and hauled them to a central location where they were organized in a way that the owners could collect them. Most didn't bother to. "It was kind of eerie," Falkman said. "They would be stacked maybe the length of a football field forty feet high, rows upon rows upon rows upon rows."

It was hard work, seven days a week. Team members found time to be off on Saturdays, spending time with their families or for recreation, maybe a dip in the pool. They found ways to lift their spirits. Slocum became one of the camp's better winemakers. Alcohol was prohibited, so some team members buried wine in the landfill and dug it up when the need arose. They obtained some from the US embassy. Slocum cleverly acquired two bottles of costly gin, which lasted almost three years. There was an active market for scotch. Food easily available back home was also hard to come by. Slocum once paid $50 for a pound of banned sausage to make a pizza, he said.

Abiding by the Saudi rules was critical, though. One evening, Slocum, after celebrating receipt of a Jeddah payment and wiring it back to the States, was returning to the camp when he stopped at an intersection to yield to another driver. The light turned red. A policeman on a motorcycle told him he had violated the law and took him to jail. At the jail, a Saudi he knew who spoke English advised him to go back and get his car, and he'd only be given a ticket. When Slocum returned with his car, the Saudi was gone.

The motorcycle cop, now angry that Slocum had left the jail, told him to take his car back to where it was first parked and return. He did so. Slocum then called two Lebanese contractors at the camp who worked for him and asked them to retrieve him from the police. He was placed in a tight cell with eleven people, his feet touching other prisoners. A day passed and he made another jailhouse call to the States to report the wiring of the payment and alert the office of his predicament. That person on the other line got distracted and failed to tell anyone. Finally, one of the two Lebanese contractors alerted somebody at the camp, who retrieved him. He'd spent two sore days in his jail cell.

The company's presence in Jeddah made a difference though, Falkman remembered. During start-up, the managers celebrated in an open square adjacent to their quarters. Their food had to be shielded from swarming flies. Diners quickly covered their plates. Flies attacked as they ate. A year later, a similar party was held to celebrate the operation's first anniversary. "Not one bug, not one fly, nothing," Falkman recalled. Twelve months of cleaning the city, of waste collection, and pest control had improved conditions. "We were collecting the waste from all the restaurants almost immediately and from the markets," he said. "The fly population was just completely gone. And for me, that was my reward for what we had done."

Jeddah officials and the city's government were also pleased.

Mayor Farsi, a great lover of public art, was committed to the beautification of Jeddah and filled the city with art. While meeting Buntrock during one of the executive's trips, he escorted him into a large room packed with miniature pieces of art and books with art designs from all over the world. "Just pick one," the mayor directed Buntrock. There were no price tags. Farsi issued the same request to all the city's contractors. Buntrock asked if he could bring a copy of the gleaming sculpture called *Peace* that stood in front of Waste Management's Oak Brook headquarters on his next visit, but he expressed concern because the piece was the creation of the artist Leonardo Nierman, a Mexican Jew. "The Arabs loved the Jews," Farsi said. How large was the company's piece? Farsi inquired. Once he learned, he said, "Make ours two feet taller." "So we did," Buntrock said. "The Jeddah Peace statue was given a beautiful spot on the Central Corniche running along the Red Sea." Placed there in 1984, the sculpture was among the first of hundreds of works by world-famous artists to be donated to the city.

Three years later, Jeddah's officials, appreciative of the company's work, awarded ACE a five-year contract renewal.

BIDDING IN BRISBANE

Two years after the Jeddah project began, Brisbane beckoned. It was 1983.

The city of more than a million residents had advertised a tender for a private company to collect garbage for all of its households. The Waste Management team pounced on the opportunity, but it wouldn't be easy. It was the first time the city, which had done the collection, would use "wheely bins," an outdoor garbage container with wheels that was less labor-intensive and used throughout Australia but unknown in Brisbane. The city also relied on one driver and two helpers, all of whom were union members, to collect the waste.

In its bid, the team proposed using a driver and lone helper to serve 750 households—which would mean eliminating one position.

Knowing job loss would not sit well with the unions, the company's labor relations executives spent time communicating with the union's leader. Dave Coleman, the labor relations vice president, met with the Australian and offered to fly him to the States to meet with unions representing Waste Management's workers. They were working hard to spin a positive story.

Waste Management won the bid, the team's price lower than that of its competitor. They had just six months before start-up. The competitor, however, had the union's support and filed an injunction against the city right before the Christmas holidays saying it had failed to follow its own procurement procedures. The city's officials wanted to avoid a battle with the union and asked Waste Management to amend its proposal to use a three-man crew before awarding the contract. The team offered a new price, which still beat the competitor's, but it was too late; the city's rules had been violated.

Even so, Brisbane's officials were eager to move forward. Almost immediately, a second tender was put out, and Waste Management won the seven-year contract, this time proposing a three-man collection team. But with union and competitor opposition, the contract would be for only half the city.

After operating for seven years with three men, Brisbane's city officials recognized the need for greater productivity. They knew cities across Australia were using wheely bins and sideloading trucks with drivers that served thousands more residents than they were.

In 1990, Brisbane sought bids again for a longer, more complicated contract that required the winner to manage the collected waste either through landfilling, incineration, recycling, or other means. The city wanted assurances that the winner could get whatever permits were needed. The Waste Management team found an ideal disposal site, an old red-clay quarry and brickworks that had been closed and was located in a sparsely populated area. The owner was willing to sell. The price: $5 million.

"If we want to grow here and have a long-term presence, we need to be the successful contractor. We need to buy this," Falkman told his bosses. They agreed. Waste Management's team bought the site, known as the Rochedale Landfill, within the city limits. Its competitors' sites were outside the city, their permitting uncertain.

Brisbane awarded Waste Management the $7-million-a-year contract in 1990. Approximately 140 days later, the company's 150-man workforce and the new 40-truck fleet were serving 137,000 households. They distributed 300,000 wheely bins.

EXPANDING TO NEW ZEALAND

By now, the company had a management team in place in the country. It was led by Bob Van Tholen and Joe Parziale, Buenos Aires mobilization veterans. Van Tholen served as country manager of Australia and had the assistance of Parziale, who led special projects. Rich Jeffrey later succeeded Van Tholen as country manager. They followed the company playbook: gain entry, start expanding. They acquired a business in Sydney called Tiger Waste and changed the name to Pacific Waste Management. The business grew its presence in Sydney, Melbourne, the capital Canberra, and Adelaide, and rapidly become one of the largest operators in the country.

The search for opportunities never ended. During a stopover in New Zealand, Jeffrey identified an opportunity at a publicly-traded business coincidentally named Waste Management of New Zealand Ltd., the nation's largest waste company with operations in Auckland, Christchurch, Hamilton, and Wellington. Weinert, Falkman, and Jeffrey, who had by then succeeded Bob Van Tholen as the Australia country manager, were soon in talks with Alton Jamieson, the managing director of the company.

They learned that two of its majority shareholders wanted to take some cash out of their holdings in a private transaction so as not to affect the price of shares trading on the New Zealand Stock Exchange. Weinert, Falkman, and Jeffrey negotiated a deal to buy a minority interest and later purchased more shares from the two shareholders to become the company's majority owner while still having other public shareholders.

This move ensured that Waste Management of New Zealand still had public shareholders, still had local ownership, the value of which they knew very well. "It gave us a presence there with a company that had our name, and we felt comfortable with the management and the owners," Falkman said. "And we thought it was a good way to get an understanding of the market in a fairly low-risk basis."

They got the company in 1987 and, with it, Jamieson, who would eventually head Pacific Waste Management. About the same time, Waste Management entered Hawaii through Oahu Waste Systems. They would soon head to the Far East, in Hong Kong and Indonesia. They had ambitions. Hawaii was the perfect place to stop on the way.

TEAM MEMBERS LISTED IN THIS CHAPTER

Steve Bergerson	Don Flynn
Jodie Bernstein	Hal Gershowitz
Dean Buntrock	Peter Huizenga
Brian Clarke	Frank Krohn
Phil Comella	Alyse Lasser
Jeff Diver	Phil Rooney
Ed Kenney	

14

A DIFFICULT 1983

Waste Management was on a tear as 1983 began. Its revenues exceeded $1 billion, an amazingly long way from its public start a dozen years earlier. It had successfully built outposts on three continents and continued its rapid expansion across the US. Everywhere it went, growth followed. But that growth spawned problems. It meant a quick entrance into more challenging hazardous waste service businesses. And at times, it also meant some shortcomings in controls and oversight in these new businesses.

Before the year's first quarter passed, the company's success would make it a target as competitors, governmental agencies, and investigative media focused on its operations and shortcomings. And there was one place where no company wants its problems raised.

On March 21, 1983, *The New York Times* had Waste Management in its gunsights. Its front page read: "Giant Waste Company Accused of Illegal Acts."

The long piece by investigative journalists Ralph Blumenthal and Raymond Bonner reported an array of environmental issues facing the company. *The Times* had caught the company off guard. Waste Management had become aware of the newspaper's inquiry only a few days earlier. Broadly, the story included allegations of noncompliance at disposal sites with federal and state hazardous waste management rules in a number of states, environmental fines, and accusations of antitrust violations. The publicity was a major blow to its reputation and to its standing on Wall Street, which had held it in high regard since its 1971 public offering. In the immediate aftermath of *The Times'* story, the value of the company's stock approximately halved. Its stock was the most actively traded on the New York Stock Exchange three days in a row.

The Times' coverage was the company's first crisis. Waste Management's leadership took immediate action. They announced the company would hire an outside legal counsel to undertake an investigation. Hal Gershowitz took the lead in assembling the response team, and Steve Bergerson, the company's general counsel, brought in lawyers. They would hire experts in environmental law and compliance and governmental affairs. They would be people who had credibility and credentials.

THE ENVIRONMENTAL LAW EXPERT

Joan "Jodie" Zeldes Bernstein was one of the first experts they contacted. Bernstein was from Galesburg, Illinois, where her father, who had emigrated from Ukraine at age thirteen, ran a tavern on Main Street, Louie's Liquors. Her mother Sophie, born in Poland, was a department store buyer. The eastern European immigrants impressed upon their children the value of education. During the 1930s and 1940s, the Zeldes was one of only twenty Jewish families living in Galesburg, an environment rich in achievement. Bernstein's aunt was the nation's first woman to be a certified pediatrician. Carl Sandburg's brother, a historian, lived just two doors away. Bernstein's good friend Barbara Anders married an electrical engineer named Jack Kilby, who would share in winning the 2000 Nobel Prize for Physics for his work on the integrated circuit.

Bernstein was editor of her high school yearbook and went on to the University of Wisconsin, a "more progressive school where there would not be as much pressure to join a Jewish sorority," she recalled. She majored in economics and political science. She was on the student council and voted outstanding senior woman. A professor suggested she pursue a doctorate in political science, but she had law school in mind. "I think you should go to Yale," she recalls her professor, David Feldman, telling her. "It's different than all the law schools. All the other law schools are trade schools. With my recommendations, you'll get in." And, of course, she did, and she excelled, graduating at the top of her class in 1951. She was among eleven women in a class of two hundred.

Despite being told the big New York law firms didn't hire Jews or women, she gained an offer from Sherman and Sterling, a highly respected old-line Wall Street firm, and passed the New York state bar exam. She worked there for three years, when she left to marry a young doctor, Lionel Bernstein, and moved to Chicago. She was soon practicing law at Chicago's Schiff, Hardin & Waite. But her husband's work took them across the country, to medical positions in Colorado, California, and then back east to Washington, DC. She worked as a stay-at-home mother and part-time lawyer and occasionally would connect with her Yale Law School friend, Pat McGowan Wald, who later became chief judge of the US Court of Appeals for the District of Columbia. Wald's husband, Bob, also a lawyer, helped Bernstein get an entry-level position at the Federal Trade Commission, where she sued toy companies, won, and was noticed by important people.

After Jimmy Carter was elected in 1976, Bernstein was named general counsel of the EPA. Carter had pledged "to make opportunities for qualified women in jobs they never had before," Bernstein recalled. After Carter was defeated by Ronald Reagan in 1980, Bernstein joined Washington's Wald, Harkrader & Ross law firm, specializing in environmental matters. That's where she got the Waste Management call.

The Times' coverage had triggered fire alarms, and Waste Management needed help fast. Bergerson, the company's general counsel, called Bernstein, who, in turn, called Angus Macbeth, who had helped shape the nation's first environmental rules following the establishment of the EPA in 1970. He and seven Yale Law School classmates had organized the pro-environment Natural Resources Defense Council in 1970. And, not unimportant to Waste Management, he had been chief of environmental enforcement in the Justice Department during the Carter years. Bernstein and Macbeth were a powerful pair.

The two were on a plane to Waste Management's headquarters in Oak Brook immediately. They met in the boardroom with the company's leaders—Buntrock, Rooney, Gershowitz, Flynn, Peter Huizenga, and Jeff Diver, an in-house environmental attorney looking after the hazardous waste business. The discussion went on for hours. Bernstein had read *The Times'* articles and believed the company would soon be in litigation all over the country.

With Gershowitz's aid, they gathered files of background information and prepared a press release that responded to all the claims at all the facilities *The Times'* reporters had covered. They had worked until 3 a.m., filling the release with information about the facilities and their permits. Armed with the information, they held a press conference the following morning at a downtown hotel with Gershowitz doing most of the talking and handling the questions. The press coverage that followed was fair. The team then headed to New York the next morning to meet with the financial community and work to stabilize Waste Management's stock. Over the next few weeks, things began to settle down. Bernstein and Macbeth returned to Washington, DC, but they continued to help out.

"I know I kept four or five associates busy," Bernstein remembered. "For the first time, someone had to go check on the permits at Emelle (one of the company's Chemical Waste Management facilities that *The Times* cited). We had to go to every one of the Chem Waste sites. We had some very smart and capable young lawyers. And I sent them out to make a record. So when the state or feds got worked up, we would be able to defend."

Meanwhile, there was concern at Bernstein's Washington law firm that the environmental lawyers were going to bolt. Bernstein didn't trust the uncertain situation and mentioned to Bergerson that she was thinking of moving on and taking the Waste Management business to a new firm. She wanted Bergerson's reaction.

"Would you think of coming here?" he asked. Bernstein was soon in Oak Brook lunching with Bergerson and Frank Krohn, a talented, business-savvy staff attorney. They wanted to know whether she would be interested in becoming Chemical Waste Management's general counsel. She returned to Washington, DC, got her husband's agreement, and moved to Chicago. For a time, she lived alone in a motel room and then in a furnished company apartment. Her husband followed soon, and they bought a new home. Bernstein proceeded to strengthen the Chemical Waste Management legal staff. New attorneys arrived, among them Ed Kenney, Phil Comella, Brian Clarke, and Alyse Lasser. She also engaged outside law firms to handle local matters affecting operating facilities.

In 1983, Bernstein was the most senior woman in the company. Over time, at Buntrock and Rooney's urging, she took on the task of developing the company's business ethics program that mandated employee training and included material on sexual harassment issues. In 1996, Bernstein departed Waste Management to return to the federal government. She was appointed director of the Consumer Protection Bureau of the Federal Trade Commission.

REBUILDING TRUST

The negative press coverage prompted Gershowitz and Flynn to undertake an emergency financial tour to reach out to the financial community in the US and Europe. The company had faced a beating and needed to respond. The news had been "a bludgeoning," Gershowitz remembered. But the two men repeatedly heard from investors that Waste Management was the first company to come calling when the news was terrible. "Everyone came when the news was great. We came when the news was terrible, we were constantly told," Gershowitz said.

Their visits helped. Within six weeks, the stock reached its then all-time high.

While the news coverage's resulting picture raised broad suspicions that negatively affected the company's reputation and its management, it also led Waste Management to establish a potent force of talent in legislative and regulatory affairs in Washington, DC and in state capitals, an impressive ethics and environmental compliance program, and an effort to burnish its image among its employees, customers, government agencies, environmental groups, the public at large, and, of course, Wall Street.

It wasn't image-building just for gloss. The company was maturing. It needed new and additional internal and external resources to address deficiencies and strengthen its operational controls. And, as it grew and more issues surfaced across its horizons, it needed to better anticipate these problems and respond to the varied issues it faced.

TEAM MEMBERS LISTED IN THIS CHAPTER

Lee Addleman

Jim Banks

Walt Barber

Charles Bayley

Larry Beck

Jodie Bernstein

Calvin Booker

Sue Briggum

Dr. William Y. Brown

Dean Buntrock

Lisa Disbrow

Joe Donovan

Bob Eisenbud

Kim Engle

Will Flower

Hal Gershowitz

Bernie Handley

Leah Haygood

Pat Haynes

Peter Hendrick

Ron Hogan

Kevin Igli

Keith Jackson

Kevin Joyce

Lisa Kardel

Peter Kelly

Lynda Long

Veronica Lynch

Dave McDermitt

Chuck McDermott

Frank Moore

Lynn Morgan

Bob Morris

Jim Nelson

Rosemarie Nuzzo

James Davis Range

Debra Romanello

Phil Rooney

Mary Ryan

Jack J. Schramm

Lizzie Schueler

Jane Seigler

Ed Skernolis

Alan Stalvey

Kent Stoddard

Patti Summerville

Dave Tooley

Kathy Trent

Dr. George Vandervelde

Chuck White

Gary Williams

Peter Yaffe

Linda Zanolotti

15

WASTE MANAGEMENT GOES TO
WASHINGTON

Waste Management had taken it on the chin after *The New York Times'* stories ran in 1983. The investigation surfaced issues the company needed to address, and it also exposed its shortcomings in reaching key constituents. There were questions from Wall Street and its investors, environmental groups, customers and service communities, the general public, and even its own employees. At the top of this list were politicians and regulators—the powerful in Washington and the nation's statehouses who ruled the environmental industry. Waste Management had neglected building key relationships with them. That would not do.

After the EPA was established in 1970, environmental regulation began to ramp up. Fifteen governmental departments involving four agencies formed the agency, merging teams of new people who had operated under different systems and leadership and now, together, focused on the environment. The agency had enacted the Clean Air Act of 1970 and the Clean Water Act of 1972, which regulated emissions and discharges into the nation's air and water, and the Comprehensive Environmental Response, Compensation, and Liability Act of 1980, commonly called Superfund, which addressed the cleanup of contaminated hazardous waste sites. As the 1980s arrived, the Reagan

Administration was active in planning new and stricter environmental rules stemming from the federal Resource Conservation and Recovery Act of 1976, which governed the control of solid and hazardous wastes. The laws for solid and hazardous waste management were being updated and others being contemplated. All of these laws and regulations, it seemed, would affect the company.

Waste Management needed to stay on top of the latest. Hal Gershowitz, with responsibility for government relations, knew he needed to field a team in Washington and began recruiting throughout the 1980s.

THE DC TEAM: THE REGULATOR

A Missouri lawyer named Jack J. Schramm was the first to arrive to lead Waste Management's Washington presence in 1981. His credentials fit: He was a politician and an Army veteran. He had been a member of the state legislature and was named "one of a new breed" of state officials by the *Wall Street Journal* in a front-page story and "one of the most effective legislators ever to sit in the Missouri House" by the *St. Louis Post-Dispatch*. He had run for state office several times. A Democrat, he had sought the lieutenant governor's job but lost when Republican Richard Nixon trounced Democrat George McGovern in the 1972 presidential race. Schramm then ran for Congress in 1976, losing by the narrowest of margins. He cared about policy and the law, and he had a history of introducing pro-environment programs.

"The environment's protection was just getting started," he said of the 1970s. "Some of the states weren't doing enough. They wanted to pass weak environmental laws. I had good bona fides, and I was invited to join the Carter Administration at the EPA."

Schramm became administrator of the EPA's Region III in Philadelphia in the Carter Administration, where he was responsible for five Mid-Atlantic states. He was a regulator with a reputation for toughness. He faced recalcitrants in the waste industry who were unhappy with the tougher rules and personally negotiated nationally significant cases. He was respected by both Republican and Democratic governors, which earned him an offer to become the EPA's enforcement chief, which he declined.

"I was the last appointee of the Carter Administration regional administrators to leave," he recalled. "And it was because I had several Republican governors, and I always treated them well. I didn't play politics with them, and they weren't eager for me to leave."

The same year Ronald Reagan and his rebounding conservatism entered the White House, Schramm joined Waste Management. He found an office off Pennsylvania Avenue, hired an assistant, and then began the job of opening the right doors and adding the right staff members. The company was ready to have full-time representation in Washington.

"Waste Management was growing by leaps and bounds," Schramm recalled. "I had both the political background and the environmental background." He soon developed a working relationship with Anne Gorsuch, appointed by Reagan as EPA administrator. Schramm filled her in on how an EPA region worked. Schramm and Gorsuch, a former member of the Colorado legislature, had similar career backgrounds. Under Reagan, Gorsuch began to back away from new regulations and resources, which antagonized critics, Schramm said. Gorsuch was soon out, and Reagan named former Deputy Attorney General Bill Ruckelshaus, who would later lead Waste Management's competitor, BFI, to replace her. With Ruckelshaus in place, things quickly calmed down.

THE JIMMY CARTER REFERENCE

Gershowitz also needed a leader to expand the company's governmental outreach. He found him in Frank Moore, who would greatly expand the company's federal and state government affairs outreach. Moore was a well-known operator in Democratic circles. He had been President Carter's congressional liaison and had a long history with Carter dating back to early in his career and later in the Georgia capital. He had impeccable political credentials and was wise in the ways of Washington, DC and state capitals. He talked the talk of politicians.

Moore had grown up in Dahlonega, Georgia, a town of about 2,500 whose name is Cherokee and means "yellow money," he says. It was the site of the nation's first gold rush in 1829, and the town enjoyed a gold boom in the 1830s before the 49ers moved on to new riches in California. Moore majored in business at the University of Georgia. There, *The Washington Post* noted in a 1977 story, he had a marvelous time as a frat man. The paper had a good source: his mother.

Moore was knocking about after college but found himself in a trainee's job at Quaker Oats in Knoxville, Tennessee. While there, he test-marketed the company's Cap'n Crunch breakfast cereal in the Knoxville area and Buffalo, where TV markets were suited for quick product analysis. He helped market cake mix and cornmeal, grits, and flour.

Before long, he took another job at the planning and development commission in Gainesville, Georgia, his wife Nancy's hometown. He began running the local Head Start program, which helps prepare children from low-income families for school. There, he met a peanut farmer from Plains, Georgia, named Jimmy Carter. Carter was heading the largest such program in central Georgia and was looking for someone to run the Middle Flint Planning and Development Commission in Ellaville, Georgia, which he chaired. After a search, Carter hired Moore as his executive director. Moore earned $9,500 a year.

Moore became Carter's chief of staff when Carter was elected Georgia governor from 1970 to 1974, becoming a national political figure. "I was staying home and running the salt mill," Moore said. Moore would attend governors' association meetings in Carter's place, since Carter, Moore said, thought they were useless. "And he was right, they were useless." But Moore got to know the key governors and staff—the important people in the political trenches. And as so many presidential wannabes made their way to Alabama to see Gov. George Wallace, who was then running for the Democratic presidential nomination, they'd stop in Atlanta and visit Carter and his staff.

It wasn't long before Carter realized he was just as able—if not more so—than this group of traveling politicos. When Carter decided to run for president in 1976, Moore raised $5,000 to send out the announcement. He had the contacts and became Carter's national finance chairman and then deputy campaign director. When Carter won, Moore went to the White House and served as his special liaison to Congress for four years. He worked to turn Carter's ideas into law.

After Carter's defeat in 1980 and a stint at Harvard's advanced management program ("I changed my vocabulary and brushed up on what I never learned or had forgotten, you know, at the business school in Georgia"), Moore moved to Houston where lots of people were getting wealthy on high-priced oil. When the boom busted—or, as Moore says, it "didn't hit a bump in the road, it hit into a canyon and spilled off a big cliff"—he changed course. With oil in rapid decline and the future uncertain, Moore met with Charles Kirbo, an influential Atlanta lawyer and one of Carter's closest advisors. Kirbo suggested that they walk down the hall from his office to visit Griffin Bell, the attorney general during Carter's presidency. Moore and Bell knew each other well from the Carter years. Bell had even recommended Moore to Harvard.

Kirbo recalled that Bell had met someone influential duck hunting. "'Griffin, who was that guy you were duck hunting with over in South Carolina?'" Moore recalled Kirbo saying. "Oh, hell, what was the guy's name that owned the garbage company over there? Wasn't there a guy there from Waste Management and that they've been told to find somebody to set up a government affairs operation?"

The man from South Carolina was George Dean Johnson, a Spartanburg lawyer, and a close friend and business associate of theirs. They called Johnson.

Moore went to the Kentucky Derby every year as the houseguest of the mayor of Louisville. While there, he got a call from Johnson, who said, "I'm calling on behalf of Dean Buntrock. Look, can you come to Chicago tomorrow?" Moore recalled. "I said, 'Well I'm here with fifteen governors,' and I was, fifteen Democratic governors. 'And I'm having dinner with them tomorrow night and all their spouses. I'm the only civilian there, me and the mayor of Louisville.' So, I was saying that as an excuse, but Johnson took it as, 'Shit, if you know that many governors, you must be the kind of guy we're looking for.'"

Overnight, Moore became a candidate for Waste Management's governmental affairs operations. He knew almost nothing about the company and had to decide whether to go to Chicago for a visit. He did his due diligence. The White House always had an FBI agent handy. Bell connected Moore with an unnamed FBI agent to learn more about the company.

"'Give me twenty-four hours,'" Moore said the FBI agent told him. "He called me back and said, 'I cannot give you the file, and I can't tell you what all is in it.' It was three inches thick. 'They've been investigated and investigated and a summary sheet on the top said that these people, this company, is run by Dutchmen, and they are the only ethnic group that is stubborn enough to keep the mafia out of their business.' He said they're clean."

Moore then contacted bankers A.D. Frazier at the First National Bank of Chicago and Charles Duncan at the Bank of Commerce in Houston. They both responded that Waste Management was a reputable company and their banks wanted to do business with it. "While they were checking me out, I was checking them out," Moore said. Moore had Jimmy Carter as a reference and a recommendation from Bell. Buntrock and Gershowitz decided to call Jimmy Carter. Buntrock's administrative aide, Rosemarie Nuzzo, dialed the former president's number and quickly got past his gatekeeper. Buntrock reported him saying, "'Hello, Jimmy Carter.'" It was a very good reference. President Carter liked Moore a lot.

MOORE BUILDS GOVERNMENTAL AFFAIRS

Moore joined the company in 1983 as vice president of government affairs. And just as fast, Beck had him on a jet traveling the country visiting its facilities and getting introduced to the company's senior operators. The operators were skeptical of him, Moore thought. They had seen a train of corporate recruits pass by before, but Moore was determined to gain their trust.

Building the governmental affairs operations was imperative. Moore instinctively knew the company needed good government affairs people in the regions, closer to the operations in the key states. The federal government was moving fast to stricter regulations and delegating its enforcement authority to state environmental agencies. The company needed respected and informed team members engaged with the government and monitoring it for rule-making that could adversely affect its operations, or, just as important, advantage the company.

The company's operations were organized into a network of geographic regions each overseen by a region manager. Moore recruited government and regulatory affairs persons and assigned them to each region. The managers paid for them, absorbed their expenses, and furnished their cars, but they reported to Moore on the dotted line. A small number assisted from the Oak Brook headquarters.

His goal was to introduce the company to both Washington and the state capitals. Many public officials only knew the company by what they read in the papers. Moore installed staff in state government affairs positions and developed a national network of lawyers and lobbyists.

"I was teaching government leaders about Waste Management and Waste Management about government. Working the government wasn't a separate business, it was part of the business," he said. "You had to have good relations. You were always hitting bumps. You were always making mistakes. You'd have to go in and be able to talk to people and mitigate it. Or kill a bill here, kill a big bill there, or take a bill and manage it for two to three years. These things would all come in big waves. They'd introduce some [bills] in California, the next year it would be in Wisconsin. So, we'd have to develop all the talking points, all the white papers, and we would have to educate our people on how to complete those and go into their respective legislatures."

The company also developed relationships with influential law firms that had government affairs experts. Moore understood the value of bipartisanship and how elections could change influence in state houses. "They [the law firms] usually had a real respected Republican and real respected Democrat. And they were needed depending on who the governor was," Moore said. "And then we had a regulatory person, a three-legged stool." Moore proposed adding even more internal government affairs staff in Oak Brook, which was denied. They were to be in the field close to operations.

Moore later added three government affairs team members in Oak Brook: Ron Hogan was recruited from the National Conference of State Legislatures in Denver and Keith Jackson from the Illinois legislature. A third, Jim Nelson, arrived from FMC Corporation, a chemical manufacturing company, where he had directed state affairs. Their first assignments were to find key people in each state who could develop relationships with government officials and look after legislative and regulatory matters. They also worked to educate operations managers on the fundamentals of government and how it could benefit or disadvantage their businesses.

The three would help establish the government affairs network and eventually run regional government affairs offices: Hogan served in the Southeast, Jackson in the Midwest, and Nelson in the Northeast.

THE REPUBLICAN

Reagan's arrival in Washington in 1980 sharpened the company's focus on Republicans. The Democrats were out and the Republicans were in. Gershowitz and Moore knew they needed more muscle and savvy in Washington, where the environmental regulations were written and would be passed along to the states. That meant one thing: They needed a Republican.

Moore went to visit Sen. Howard Baker of Tennessee, the Senate majority leader who later became Reagan's chief of staff (and even later, a Waste Management board member). Baker worked well across the aisle with Democrats. Baker directed Moore to James Davis Range, or "Jim" as he was affectionately known. The handsome—some said blustery—Range was a prominent policy strategist and conservation advocate.

Range, who grew up in Johnson City, Tennessee, attended Tulane University as an undergrad and earned his law degree at the University of Miami. He had served as minority counsel to the Senate Committee on the Environment and Public Works from 1973 to 1980. He later became chief counsel to Sen. Baker from 1980 to 1984 when he was Senate majority leader. Range was well-connected. It seemed like everyone in Washington knew him. He was an avid hunter and fisherman, an out-doorsman who was a leader of several conservation organizations. He even raised hunting dogs.

Range had a well-earned reputation for promoting natural resource conservation. He was an original board member and chairman of the National Fish and Wildlife Foundation. He was on the boards of Trout Unlimited, Ducks Unlimited, The Wetlands America Trust, the Recreational Boating and Fishing Founding, the American Sportfishing Association, the American Bird Conservancy, the Pacific Forest Trust, and the Yellowstone Park Foundation.

Range was well-known by senators and their staff as a keen policy strategist and information source. He knew the politics and horse trading inside the capitol. He played leading roles in drafting some of the nation's most protective environmental legislation, including the wetlands provisions of the Clean Water Act of 1972.

Sen. Baker was clearly fond of his former aide, and although Range had a reputation for organizing bipartisan support for his beloved conservation bills, he was Republican to his core. "'Well, I'll tell ya, he believes that all Democrats when they are born they ought to cut their tails off and put them down a well,'" Moore recalled Baker telling him. "'That's just a problem you have to deal with if you hire Jim Range.'" Moore, a Democrat, said he'd handle it. Range became head of the company's Washington, DC office in 1982, though the Moore-Range partisan divide persisted. They were, after all, party men.

Range began by hiring a staff that could lobby the Hill, commu-nicate with the EPA and other agencies, and develop alliances with environmental groups. One after another they came: lawyers, lobbyists, regulatory experts, and recognized environmentalists such as Jim Banks, Bill Brown, Sue Briggum, Bob Eisenbud, Chuck McDermott, Jane Seigler, Ed Skernolis, Kevin Igli, Kim Engle, Leah Haygood, Lizzie Schueler, and Linda Zanolotti.

The office was collegial. As the workday ended, relaxation set in. Jokes would be told, gossip shared. Drinks were often available. Range had a taste for Jack Daniels, a Tennessee whiskey. No one paid attention to organizational charts; there was work to do.

Range brought his hunting dogs to the office, too, as he had throughout his career. Range loved his dogs. When he worked at the Senate, he had an unusual training ground: the Capitol Rotunda. He would park his pickup outside the building, bring his retrievers into the hallowed halls, and throw items every which way for the dogs to fetch.

Strategizing was Range's strength. He was creative at rulemaking. A colleague recalled that while he did not write well, he acted like a conductor in guiding people to what needed to be written and fixed. "He got the biggest picture, the macro strategy, and sense of what needed to be done," she said. "When it came to certain tactical execution, he'd outsource that to really good people in the office who could write well."

THE ENVIRONMENTALIST

Jim Banks joined Waste Management's DC office in 1985 as its director of regulatory affairs. He carried impressive credentials. Banks had directed the National Resources Defense Council's (NRDC) Project on Clean Water, leading a coalition of environmental groups to lobby the 1977 amendments to the Clean Water Act. He was at his core a conservationist and an environmentalist. He came to Washington as a staff attorney for the US Marine Mammal Commission after graduating from the University of Michigan Law School, which had an elite environmental law program. He had been the outstanding senior in his class at the University of Kansas, where he received his undergraduate degree in civil engineering.

He knew Range from his work on the Hill. Banks had wanted to switch from water issues to hazardous waste, but another NRDC person had that role, and he was stymied.

Banks generated attention when a group of environmentalists wrote a letter to *The Washington Post* attacking an EPA official for being under Waste Management's influence whom Banks knew well and respected. The move did not sit well with Banks. Banks fired off his own letter charging the group with character assassination and defending his friend. Range noticed and reached out to Banks. They started talking and were soon on their way to meet the company's leadership. He joined the DC office and said he was the first senior public interest person in history to "leave a green group and join a company. There was quite a stir about that."

During the 1980s, a number of environmental groups targeted the company. But Banks had advised the NRDC's executive director, a friend, of his move and that he'd be working on hazardous waste issues. Word spread that the move was a good one and it quieted much of the criticism. A story in the Environmental Law Institute even quoted his friend saying Banks was "an honest broker."

At the NRDC, Banks had spent his time suing the EPA to enforce Clean Water Act rules. Now, at Waste Management, he advocated for the incineration of organic hazardous wastes on land and at sea, a boon to the company's Chemical Waste Management business but also something he believed was the right thing to do. At the time, the company was anticipating new regulations for at-sea incineration and had invested in two ships, the *Vulcanus I* and *II*, to burn the material in the Gulf of Mexico. The EPA was, in fact, promoting the technology for its effectiveness. However sound, the ocean incineration program would ultimately fail after the outcry of environmental groups.

During his career, Banks focused on a range of hazardous waste issues as well as Superfund. He also responded to efforts by state and local governments to set flow-control rules designating the facilities to which solid waste would be taken. He wrote policy papers and commented on proposed environmental rules. He lobbied. And it would not be long before he found himself at the Oak Brook headquarters, succeeding Jodie Bernstein as general counsel of Chemical Waste Management.

THE SCIENTIST

Dr. William Y. Brown joined Waste Management's DC office about the same time as Banks, in 1985. And like Banks, he had been brought on to help in the battle for ocean incineration. It was "all hands on deck," Banks recalled, and the group included the captain of the *Vulcanus I*.

Waste Management needed to participate in the legislative and regulatory process, Gershowitz said, and Brown was among the heavy hitters who had impeccable credentials. "We wanted the Washington office to be among the most respected people in the environmental and legislative areas," Gershowitz said.

Brown had arrived from the Environmental Defense Fund, where he had been a senior scientist, attorney, and acting executive director. He was needed for his lobbying skills, and his presence sent the message that the company wanted to align itself with the environmental community.

Brown was thoughtful, cerebral, scientific, and, like the others, very smart. He had been executive secretary of the US Endangered Species Scientific Authority. He had played a key role in stopping the repeal of important Endangered Species Act provisions. In addition to being a Harvard-trained lawyer, he had earned a doctorate in zoology from the University of Hawaii, where he was a National Science Foundation fellow.

Brown began as a governmental affairs director and focused much of his attention on the company's environmental programs, guiding it to adopt its first comprehensive environmental policy statement, which held managers accountable for the performance of their operations, including pollution prevention goals and a program ensuring no net loss of diversity on company properties. Brown and colleague Leah Haygood developed the company's first annual environmental report, a document that reported the company's compliance efforts and record. Waste Management was among the nation's first companies to issue such a public disclosure.

He also guided the company's participation in the Environmental Grantmakers Association, an organization made up of environmental funders, and the Council of Foundations, which fosters a community of philanthropic professionals. He directed Waste Management's grant-making, providing between $500,000 and $1 million annually to a range of environmental groups.

Brown later became Waste Management's vice president for environmental planning and program and chaired the company's executive environmental committee.

CHEM WASTE'S LIAISON TO THE EPA

In the early 1980s, Kevin Igli had been working on cleaning up PCBs, or man-made chemicals that were largely banned in the US in 1979 for their toxicity, for a subcontractor at Waste Management's deep well hazardous waste facility in Vickery, Ohio. He soon got a call that changed the course of his career. Dr. George Vandervelde, a Chem Waste vice president, invited him to join his environmental management unit. Igli arrived on his twenty-fifth birthday, July 24, 1983.

His job was compliance. After *The New York Times* articles, the company needed to ensure that its rapidly growing Chem Waste unit was in compliance with environmental regulations. Igli traveled to Chem Waste facilities to make sure they were operating within the rules, wrote reports, and, if he found something amiss, ensured it was fixed. His knowledge of the rules became his calling card—and attracted attention. Bernstein, Chem Waste's general counsel, Banks, and Walt Barber, vice president of environmental management at Chem Waste, respected him and said they had an idea: Would he go to Washington and help the company respond to the changes emerging from the amendments to the federal RCRA law regulating hazardous waste facilities? He would.

Igli arrived in DC in 1986 and reported to Banks, the director of regulatory affairs. He worked on hazardous waste regulation and lobbying. He focused on hazardous waste regulations.

"My job was to go to the EPA every day and build relationships and get to know all of the branch chiefs in the RCRA office, get to know the division directors, and just get in on the inside of what was going on and how the rules were being developed," he recalled.

Igli worked the hallways and made fifteen-minute appointments with EPA staff. He sought "to represent Chemical Waste Management and Waste Management as companies that wanted to be on the forefront of doing good things to protect the environment by having sensible regulations to treat hazardous waste properly."

He began to build a relationship too with Range, the head of the Washington office, and would join him and other colleagues hunting in Maryland shore duck blinds. The hunters were often accompanied by high-level government officials Range had invited.

Range and Igli had a congenial, close relationship. Igli recalls how Range would lean against his office door and stare at him until he looked up. "He'd say, 'What the [expletive] you doing, boy?' I'd start laughing and so would he. He'd also come about his dogs. One of his Labradors took to me, and so I got the assignment of jogging with the dogs at lunchtime a lot of days. The dog's name was Plague. Range would lean on my door and say: 'Kevin, is that damn Plague under your desk?' I'd say, 'There's not a plague under my desk, but there's a dog named Plague under my desk.' He'd say, 'Well, are you going to walk that son of a bitch at lunch?' I'd say, 'Yessir.' Plague and I would go to Rock Creek Park. How do you not love a guy like that? He was just a hot mess. He was serious but funny. He was passionate. He was sometimes half crazy, you know. Golly dog, he was something else. He was just a remarkable individual in my life."

After four years in Washington, DC, Igli returned to the Oak Brook headquarters and was promoted to the job of improving Chem Waste's environmental management program. He was the department's number two and soon would lead it.

THE EPA INSIDE MAN
Waste Management's government affairs division also needed someone who knew the EPA and could focus on solid waste landfill regulations. In

1987, Ed Skernolis became the company's representative working with the EPA on solid waste landfill regulations as well as its adviser on Superfund.

Skernolis was from Hazelton, a coal mining town in Northeastern Pennsylvania, where his father owned a saloon. Skernolis earned a liberal arts degree from Penn State and served a stint in the Army in Korea. He thought he'd become a teacher but found himself taking a civil service test for a job in the federal government. He interviewed at the IRS and EPA, and ultimately opted for a job in information systems at the EPA Region III in Philadelphia in 1972. There he met "a bunch of guys in Timberland shirts with their sleeves rolled up," he said. He was among the agency's earliest outside employees.

Skernolis got to know Jack Schramm, then the EPA administrator for the region and became his assistant. When Schramm left to join Waste Management, Skernolis managed government relations work at the EPA and eventually found his way into the agency's Superfund program, investigating old hazardous waste sites. "My job was to find Superfund sites. One of the first places to look at was old landfills that were the recipients of garbage and other kinds of waste," he said.

Skernolis had interviewed with Schramm earlier, but the company was then looking for a person to focus on Congress. That changed in 1986.

"In 1986, a couple of things happened. There were amendments to the Superfund law and the solid and hazardous waste regulations," Skernolis said. "The Superfund program was going to get much bigger and aggressive. And the EPA was going to develop regulations for solid and hazardous waste. I was putting so many sites on the Superfund list that Waste Management either had to hire me or kill me," he joked.

Skernolis joined the company in January 1987 and played a key role in the company's response to the EPA's emerging Subtitle D regulations, which set standards for designing and operating municipal solid waste landfills. The rules, which would go into effect in 1991 (followed by a series of deadline extensions), aimed to establish criteria for financial assurance, remediation, and closure of solid waste disposal sites. The rulemaking was critically important to the company, which operated an enormously valuable landfill network. Meeting these regulations was going to be expensive.

Skernolis, working closely with Don Wallgren, the head of the solid waste division's environmental management department, attorney Peter Kelly, Gary Williams, Wallgren's right-hand expert on the environmental management staff, and others, developed a draft regulation they shared with the EPA. The draft, made up of fifty pages of comments, served to bring the company's operating expertise into the agency, Skernolis said. "They needed to hear about the reality of operating a municipal waste landfill in the real world," Skernolis recalled. "It's a function of communication, cooperation, expertise transfer, and sharing of ideas. It's a collaborative process."

They shared their ideas generously with the agency, which produced a draft regulation in 1989, and were generally pleased with the rules when they finally came out. Waste Management, like the other large public companies, had the financial and technical resources to respond, especially to handle the great expense of landfill liner systems.

THE LOBBYIST

Bob Eisenbud had been managing director of environmental policy at the NSWMA when he came to the company in 1988. "We stole him from the association," a colleague said. He was the company's primary lobbyist on Capitol Hill, and he was a very busy man. There was much to protect, much to influence.

Eisenbud was primarily concerned with two issues: a ban on shipments of waste across state lines and another called flow control. The former sought to disrupt age-old systems that cooperatively served communities across the country and the many landfills that hugged state lines and served multiple states. The latter, flow control, identified local governments directing the movement of wastes to select disposal sites in their jurisdictions. He handled a third legislative issue, too: tax credits for renewable energy that aided the company's landfill gas recovery projects. These credits made the landfill gas recovery systems viable and came up almost every year. Capturing the gas meant fewer greenhouse gas emissions, an environmental benefit.

Eisenbud was a master of environmental and marine law. He once had been chief counsel for maritime and oceans policy for the Senate Committee on Commerce, Science, and Transportation; general counsel to the Marine Mammal Commission; and special counsel to the Environmental Defense Fund and Greenpeace International. He checked the right boxes, and he fit right in. He joined colleagues in watching the *All My Children* TV soap. He'd go to the loading dock to smoke, chatting up locals. And it didn't hurt that he was a fly fisherman.

As a lobbyist, Eisenbud was the consummate professional: highly respected, methodical, unfailingly polite, and well recognized in the Capitol. Every member of the Senate seemed to know and admire him. More importantly, they took his calls.

What he wasn't was a hallway lobbyist. "He understood the process of legislation and the importance of phrases and words in legislation and what the implications were if you didn't do it really well," said Skernolis. Another colleague called him "surgical" for his precision in analyzing the language of legislation.

"You really needed a green-eyeshade person looking over the text of a piece of legislation to make sure you didn't get screwed by not paying attention to a word here or a word there," Skernolis said. "That was his personality and absolutely necessary for the job at hand."

Knowing the issues was critical. "The work of lobbying is not running up and chatting with your old buddy congressman so-and-so," Skernolis said. "It was months and years of work at the staff level with people who are technical experts in the field. You really needed to know your stuff to be effective. Obviously, the personal relationships helped enormously. You had guys like Jim Range and Frank Moore who took care of that. That's the phone call you needed at the end of the process. That might happen after the end of months and years of work to get something done right." Eisenbud never wasted a member's or staffer's time, government affairs colleague Sue Briggum recalled.

Interstate waste, or the transfer of garbage across state lines, was always among the office's top issues. "He alone can take credit for keeping waste and recycling as part of interstate commerce and not some inferior business subject to lesser standards under the Constitution," Briggum said. "Many pollution prevention and recycling services need a regional, not a state or local limited market, to thrive. And he was focused on getting the right allies in the right places to tell his story."

"It was a big job, and he took care of it," Skernolis said. "To this day, there are no controls on the interstate transport of garbage." For the company, no legislation equaled success.

THE LAWYER

Sue Briggum arrived at the DC office in 1987 after having already worked for the company at two outside law firms on assignments from Bernstein, Carolyn Lown, a company environmental lawyer, and Don Wallgren, the vice president for environmental management. She had talent and presence and a knowledge base the company needed.

Briggum grew up near Harrisburg, Pennsylvania, attended the University of Pittsburgh, and earned her Ph.D. in English literature at the University of Wisconsin–Madison. She had thought about becoming an attorney years earlier but was put off after a lawyer presenting at an informational session welcomed "future legal secretaries." She had larger ambitions. She received her law degree from Harvard and spent many years working in a private law practice in Washington, where she focused on environmental regulations for utilities, chemicals, oil and gas, and mining industry groups. She arrived at Waste Management already familiar with its business and experienced in the world of environmental regulations. She also knew some of the company's most important customers.

"The thought was that they wanted someone who had submitted a million sets of comments to the administrative record with EPA and DOJ," she said. "They also wanted somebody who had links to the business community because everyone in the office had either come from the Hill or an environmental organization," she said. Her legal work in Washington had connected her to people in industries confronted with increased environmental regulation.

"Waste Management's reputation in DC at the time was that they were very aggressive in seeking the most stringent possible requirements in order to profit from the infrastructure that you had to build to meet those requirements, and they weren't really listening to customer groups."

While the company understood that strict regulations advantaged it, its leaders realized that balance was needed, too. An evolution was occurring. Stricter rules meant that waste generators that had been paying $5 a ton to dispose of in the old town dump would be compelled to pay $50 to use highly regulated, highly engineered facilities equipped with liners and monitoring systems. The higher cost was needed to prevent pollution of the soil, air, and groundwater. It also generated a decent profit.

Waste Management needed to hold itself to the highest standards and insisted that other industry leaders do the same. "But at the same time," Briggum recalled, "we weren't the only smart guys in the world, and we needed to appreciate that there might be more efficient, less expensive ways to provide the equivalent protection." The company also needed to understand its customers' perspectives.

She said the leadership's direction was clear: "Environment is number one. But also, listen to your customers and respect their knowledge and see if there is a way that you can find a consensus position that both the community and business community would be relatively satisfied with and the government would agree was a balanced outcome," she said. "We saw ourselves as a fair-minded middleman between the demands of the environmental community and community members and business communities who generated waste."

Briggum remains thoughtful about the office's approach. "We wanted to earn our credibility, and our ability to make recommendations about regulatory approaches by being a trustworthy source of hard data," she said. "We tried to be transparent in the information we provided, we tried to be specific and detailed, and we thought that was the best way to convince the regulatory community that we were a voice that should be consulted.

"On the lobbying side, it was a little different," she said. "You don't go into a room with a bunch of scientists and try to convince them of something in two hours; you would have ten minutes to have an elevator talk with someone. It was extremely important to have Range and Banks and Brown who were very wired, that members of both the House and Senate knew them. It was easy to go in and talk in a broad outline. People trusted them because they knew them. They worked with them, they socialized with them for decades. That was, at the time, your classic legislative approach."

Briggum, who co-authored two books on Superfund, added: "Regulatory affairs are very, very different because even people who had run an agency can't really go back and talk to civil servants and convince them of anything unless they were trusted because of the information that they provided. Being a pleasant person never hurt. But the important thing is that you have to be reliable. Document after document, topic after topic, show up, give them information that turns out to have been fair and reliable, and eventually, you build your reputation that way. It's not glamorous work for an environmental lawyer, but it's really interesting because it's very specific, detailed, and technical."

Briggum worked with other colleagues, including Range and Skernolis, on crucial issues such as Superfund, environmental justice, and a broad range of regulatory matters. Range would sometimes jump into action on regulatory affairs, sometimes frightening Briggum. She recalled that she might visit Range to discuss a policy position and expect to get his feedback and counsel in order to refine an argument. Instead, he'd act. "Too many times he would just pick up the phone and call that chairman of the Senate committee and tell him the idea, and it was terrifying."

Building allies was important. Briggum was appointed to a dozen federal advisory committees, each composed of members from industry, state and local government, environmental groups, tribes, community groups, and academia who addressed environmental topics. Briggum found the groups were a helpful way to network in the industry and build credibility. She found herself being helpful ("my favorite thing to do") and drafting solutions to issues. She also served on the Superfund Action Coalition; the National Commission on Superfund; the Enterprise for the Environment; and the President's Export Council.

She had an influential thirty-three-year career, retiring in early 2020.

THE ENVIRONMENTAL CRUSADER

Chuck McDermott, who would eventually lead the Washington office, had politics in his blood and rock music in his soul. He was the son of a successful Dubuque, Iowa, lawyer who had worked for John F. Kennedy's 1960 presidential campaign. And he grew up playing guitar—a hobby that would fuel his early career.

Kennedy's victory brought the McDermott Irish Catholic family to Washington, where his father worked on the National Security Council during the Cuban Missile Crisis. Attending a local parochial school, McDermott's classmate and friend was Joe Kennedy, the eldest child of Robert Kennedy. He spent hours at the Kennedy's Hickory Hill home in McClean, Virginia, where one might find astronaut and later Sen. John Glenn, singer Andy Williams, and ballet dancer Rudolf Nureyev hanging out. There were touch football and ski trips and river trips and a close friendship. When Robert Kennedy was assassinated in 1968, McDermott was an altar boy at his funeral in New York's St. Patrick's Cathedral. He then spent the summer with Joe in Spain, an effort to engage the young Kennedy in a distraction away from Washington.

McDermott grew up immersed in government. "It was in the air," he said. "You breathed all of that." He moved from the Jesuit Georgetown Prep to Phillip's Academy in Andover, Massachusetts, and then on to Yale, angering his parents when he decided to take a year's leave after his sophomore year.

He started playing the guitar at age nine. McDermott's year leave would take him on the road with his music for fourteen years. His band created two albums, reviewed by *Rolling Stone* and *The New York Times*. He was in California making music when he got a call from his friend Joe to work on Ted Kennedy's 1980 presidential campaign in Iowa. The opportunity was like earning a doctorate in politics, and McDermott signed on. After the campaign, he returned to California, got married, had a child, and "needed a job that provided health insurance," he said. After entreaties from his friend Joe, he decided to join the Citizens Energy Corp. in Boston, a nonprofit company the Kennedys backed to help people pay their electric bills. In 1986, when Tip O'Neil, the Boston Speaker of the House, decided to retire, Joe won the seat. McDermott managed his campaign and served as his chief of staff for two terms. Afterward, he was restless.

That would change with a phone call. In 1989, he got a call from Tom Nides, a top aide to House Speaker Tom Foley. "'I know you're looking for something. There's a guy you should meet. He's a good friend of the Speaker's,'" McDermott recalls Nides saying. The guy was Frank Moore, then Jimmy Carter's legislative director. McDermott met Moore and then Range. Waste Management's Washington office needed more political balance, that is, a Democrat. Range didn't appear eager to bring on another Democrat. But a call from Phil Rooney to Range to do so sealed the deal. They had their Democrat.

McDermott soon had a hot issue on his hands: environmental racism. The Commission for Racial Justice of the United Church of Christ, led by Rev. Ben Chavis and aided by Rev. Jesse Jackson, was planning a "toxic tour" across the South to publicize a report on civil rights and pollution. The report concluded that race was a factor in the location of commercial hazardous waste facilities in the United States.

McDermott was asked to look into the report. "'You're a Kennedy guy and you worked on civil rights,'" McDermott remembers Range telling him. "'So here, why don't you run with this one. We got to figure out what to do. We got to figure out how to respond.'"

McDermott gathered Skernolis and Briggum to address the report. Eisenbud soon joined the effort. If they could, how would they write a bill to respond? The company commissioned a study by the University of Massachusetts to analyze the locations of hazardous waste sites. They soon learned that only three percent of hazardous waste generated in the country was handled off-site; 97 percent was handled on-site by a generator. They drafted a mock bill, which would direct the EPA to identify the nation's 100 most highly affected areas, using air, land, and water pollution data along with demographic information. McDermott traveled to Cleveland to meet with Rev. Chavis. After a good discussion, they agreed to pursue a bill.

Rev. Chavis flew to Washington, DC (McDermott met him at the airport in his Nissan Sentra), where he met with political players and Waste Management staff and decided that Rep. John Lewis should be the bill's lead sponsor. McDermott knew Lewis. He was elected to Congress the same year as Joe Kennedy.

On the Senate side, they asked Al Gore of Tennessee, who was about to head South to a United Nations climate meeting in Rio de Janeiro, to lead the legislation. Gore needed an issue, some ammunition with which to respond to charges from emerging nations that the US was not doing enough to address pollution prevention for the poor. A program—and legislation—to address environmental and social justice was just what Gore needed. Though the legislation eventually failed in the Senate, President Bill Clinton issued an executive order in 1994 embodying its key principles.

In 1990, when Range moved over to the company's Rust International engineering business, McDermott was promoted to head the office. He stayed until 1998 when he left to join a Kennedy-backed fund in Boston that invested in environmental projects. Upon his departure, Briggum picked up the important work of environmental justice.

THE STATE GOVERNMENT TEAM

As the Washington office powered up, so did Moore's state teams. Moore hired staff members with strong backgrounds in legislation, politics, and community relations to join over time as issues arose. They worked in the state houses, in the regulatory offices, in the city and village council meetings, and in church halls and other venues loaned out for community meetings. They were a busy bunch.

Among the state government affairs hires were Kent Stoddard in California and his regulatory partner, Chuck White. The pair's new ideas in legislation earned their office recognition as one of the best government affairs teams in the state. In the East, there were Peter Hendrick, Dave Tooley, Peter Yaffe, and Jim Nelson; in the Mideast, Kathy Trent; in the Midwest, Veronica Lynch, Lynn Morgan, Lisa Disbrow, and Kevin Joyce; and in the South, Calvin Booker. In Florida, Debra Romanello, a former chief of staff for the Florida Senate, was general counsel and directed government affairs. Charles Bayley was in the mountain states. And they directed many lobbyists.

The state government affairs staff often wore two hats, a necessity required for facility expansions and development projects that almost always drew not-in-my-backyard opposition and challenges from competitors. Prominent among them were Bob Morris in California, Mary Ryan in Chicago, Bernie Handley for Chem Waste in Chicago, Will Flower in Illinois and later New York City, Lisa Kardel in the Mid-Atlantic states, Calvin Booker in the South, Patti Summerville in California, Dave McDermitt in the Northeast, and Pat Haynes in the Southwest. Alan Stalvey, based in South Carolina, oversaw the company's affairs in Columbia, including its low-level nuclear assets. He would later lead the Washington, DC office. The list of people who engaged in community relations, such as Joe Donovan from Oak Brook and Lynda Long in Florida, was long. They were needed to generate support for the many projects. Their work often decided whether a project would live or die.

Many of the state staff were astute development people, who mixed their backgrounds in and knowledge of the industry with a keen sense of business, politics, and company purpose. Lee Addleman in the Midwest stood out among them. A Vietnam veteran who was an advanced scout in the 82nd Airborne unit, Addleman had started out on a truck hauling garbage, or "G," as he called it. He had worked in sales and on numerous landfill projects. He got to know intimately every county or village public official, administrator, and community member who had a role in deciding whether a landfill would be sited or expanded. "An expansion was only good if both the county and company benefitted fairly," he would say. Public officials respected him, and he was always welcome in their offices.

More people would be added as the company grew—at Chemical Waste Management to navigate its complicated issues and at Waste Management International as it spread across the globe—which, in the next few years, it would, quickly—and at Wheelabrator Technologies, which would become a majority-owned subsidiary and needed a voice for its waste-to-energy plants. The government affairs and community relations managers were the public face of the company.

TEAM MEMBERS LISTED IN THIS CHAPTER

Dean Buntrock	Fulvio Guidotti
Ron Broglio	Tom Hau
Siuwang Chu	Joe Holsten
Mike Collier	Wayne Huizenga
Bob Coyle	Bill Keightley
Ronan Dunne	Jim Koenig
Ed Falkman	Rich Lauck
Don Flynn	Mike Rogan
Bo Gabrielson	Phil Rooney
Sinon Galvin	Judy Rosenbeck
Herb Getz	Frank Schroeder
Jerry Girsch	Tom Smith
Lawrence Glascott	Jim Wegner
John Goody	Fred Weinert
Bill Grube	Nigel Wilson

16

WASTE MANAGEMENT GOES
TO EUROPE

Up until the mid-1980s, Waste Management wasn't ready for more ambitious global pursuits. It had demonstrated it could operate beyond North America, winning cash-rich tenders in Saudi Arabia and South America that enriched its balance sheet. But only a small amount of its treasured capital had flown overseas. Its investments had been in North American growth.

In July 1987, that changed with a single act, opening the door to massive expansion in Europe. The Single European Act, a product of the European Economic Community, went into effect. Signed in Luxembourg a year earlier, the act removed trade barriers and created the world's largest trading zone, signaling unity and regional harmonization on the continent.

To the Waste Management executives, it signaled something else, too: the adoption of continent-wide rules calling for higher standards of environmental protection. Individual countries had had pollution prevention regulations, but they varied in strength and enforcement. Europe would now embrace environmental rules not seen before.

Since its beginning, the company had foreseen how regulations in the US afforded it opportunity. The Resource Conservation and Recovery Act

(RCRA) and Comprehensive Environmental Response, Compensation, and Liability Act (CERCLA), for example, which created opportunities for managing solid and hazardous wastes and cleaning up hazardous waste sites, fed its growth. Rules were good. History, they believed, would repeat itself. They decided to put their capital to work in Europe and where opportunity led it overseas.

The company was not entirely unfamiliar with Europe. The international team had established a base in London since 1977, a natural stop-off on the nearly 7,000-mile journey from Chicago to Riyadh. Wayne Huizenga had also flown to Bremen, Germany, in 1980 to acquire Ocean Combustion Service BV, operator of the *Vulcanus* vessel. The boat operated out of Rotterdam burning liquid hazardous waste in the North Sea. Huizenga and the executive team hoped to bring the technology to US waters. The acquisition took just three days start to finish to complete, and they would ultimately build a second such ship. They also had acquired a US license for Denmark's Volund waste-to-energy technology. Both technologies were intended for the US market. Through projects like these—and long-term contracts in Saudi Arabia and Buenos Aires—Waste Management had earned a global reputation for taking on challenging municipal cleaning projects and had gotten to know some of Europe's industry players. But their focus then had been on tenders, not takeovers. No longer.

ROASTING WASTE IN THE NETHERLANDS

In 1989, the *Vulcanus* connection led to the acquisition of ATM, a bankrupt European company in Rotterdam that treated ships' bilge wastes. ATM had a berth on the harbor in Moerdijk, a town in southern Netherlands, offering ships easy access to its treatment services. "Barges would come in with contaminated soil, and the clean soil would go back," Fred Weinert said.

"We felt it was going to be a bigger business because regulations at sea were going to be tightened up by the maritime authorities," recalled Ed Falkman, then president of Waste Management's international business.

ATM also had a kiln that "roasted" hazardous wastes and soils, removing the volatile contaminates. ATM, with government-approved technology and a twenty- to thirty-acre site, was valuable. A Dutch state company owned 70 percent of ATM, and the remainder belonged to businessman Kees Mourik, whose company did industrial cleaning. The Waste Management executives purchased the 70 percent for less than $10 million.

The Dutch government, denizens of a country below sea level, were worried about groundwater contamination. Dirty soil was a threat. With ATM's purchase, Waste Management began cleaning 70,000 tons of contaminated soil a year. Within a decade, thanks to operating improvements, they handled 400,000 tons a year. Success bred success. ATM also won a contract to treat the nation's paint wastes.

IN PURSUIT OF ITALY

By the late 1980s, Waste Management's international team had gotten to know a pair of Italian siblings who operated IGM, a first-class waste company based in Guanzate, just south of beautiful Lake Como and north of Milan. Sergio Marinoni and Ida Marinoni ran one of the largest waste groups in Italy. "Its equipment was in top order, and its base maintenance facility was superior to anything we had in the States," Falkman said. "You could eat off the floor."

Phil Rooney, at first hesitant about the business, according to Weinert, was won over when he saw it. "You could see this was a successful company. Everything was spotless, everything was built strong," Weinert said.

The Marinoni siblings shared Waste Management's view that regulations would change the industry across Europe. It would not be unlike the American experience, they believed. The two were interested in aligning with Waste Management, and the negotiations had been friendly, with fantastic Italian luncheons in the company's dining hall. Waste Management reciprocated, inviting them to Chicago, showing them facilities, introducing them to top management, and treating them to a feast at the Michelin-starred Spiaggia on the city's Magnificent Mile.

The company even flew in Fulvio Guidotti, a financial advisor for its Argentine operations, to help liaise in Italian and translate. The Spiaggia night included Guidotti and Sergio Marinoni singing Italian songs at the restaurant. A short time later, the Marinonis returned the favor with a dinner at the famed Villa D'Este on the shores of Lake Como. But the initial talks failed to find an acceptable deal. The Marinonis were unwilling to accept any minority protection rights for Waste Management's purchase of a minority interest.

The IGM connection, however, fostered Waste Management's interest in northern Italy. The executive team believed it to be a promising and profitable territory, and so they stepped up their efforts.

Around that time, Frank Schroeder, a former Alcoa executive living in Lausanne on the shores of Lake Geneva in Switzerland, had written to Buntrock inquiring about a position at Waste Management. He liked the organization's flat structure; the decision-making distance between him and the very top was short. Weinert followed up, connecting with him during a stopover at the Zurich airport. Smooth and likable, Schroeder had a background in developing relationships with family companies and identifying good operating units. He had lived in Europe for years and knew the territory well. "He was ideally suited," Falkman said of him. An offer letter followed.

Schroeder had one mission: find acquisitions. "The whole challenge was to build Waste Management's presence in Europe," Schroeder said. "I was the acquisitions guy. I was a marketing guy." He started hunting.

Schroeder found PITEF, a waste company that served Venice and its surrounding provinces. It was owned by Stefano Gavioli, who had built a substantial company and was a brilliant marketer who invested lavishly in promotion. Waste Management purchased 100 percent of PITEF. Gavioli stayed on to manage the company once it was acquired. Again, Schroeder and his team followed the American playbook: buy company, keep management, grow.

A year later, in 1989, the Waste Management executives were making progress on their plan. While traveling in Italy, Weinert learned from Ida Marinoni that her brother, Sergio, had passed away. Weinert attended a luncheon in Milan to pay his respects and heard that Ida had no desire to run IGM. "'Come over and let's talk about selling you 100 percent of the company,'" Falkman recalled her saying. Weinert joined Falkman, who succeeded him as president of the international business, and Schroeder, to discuss the prospect. The three bought IGM.

"Once that happened, the Italian market just completely opened up," Falkman recalled. "The Italians felt, okay, Waste Management is in the country, it's better to be part of Waste Management. Within twenty-four months, we had bought seven companies."

Acquisitions included the solid waste businesses Sirtis in Novara, Sacagica and Nova Spurghi in Milan, and Saspi and Gruppo Matteini in Florence. They would also acquire the hazardous waste businesses Ecoservizi and Italrifiuti.

The Waste Management European team analyzed countries to determine where their next investment made sense. They avoided southern Italy and its reputed organized crime influence. Schroeder introduced himself to business owners at environmental fairs. "'I would like to come and see you some time,'" he told them. "They knew Waste Management and they were also aware that the EU was requiring countries to enforce the EU environmental legislation." He wined; he dined; he invited Italian delegations to visit Waste Management's US facilities. They hosted late-night feasts at the top of Chicago's one-hundred-story John Hancock Center, exhausting the Oak Brook staff invited to join in. His tactics worked: In just three years, Waste Management had quickly become the largest operator in Italy, with 3,000 employees and $500 million in sales.

EXPANSION IN WESTERN EUROPE

Waste Management's acquisitions would not be so easy in Germany or France. Many of the companies there were not as open to outside ownership.

The Americans were welcomed to talk by the German Edelhoff and Rethmann waste companies, but nothing would come of it. They eventually were able to operate in Germany by acquiring a waste-to-energy plant in Hamm that processed 1,000 tons of material a day. In Soest, they gained a collection company that served 800,00 residential and 5,000 commercial customers. Despite these efforts, Waste Management would never establish a stronghold there.

It was doubly difficult in France. The Waste Management team was playing on an unreceptive playing field. In 1987, Greenpeace distributed a report on Waste Management to members of the French parliament and to the heads of local governments. The report, part of a campaign to target the waste business and promote zero waste generation, chronicled incidents of the company's regulatory noncompliance. Company opponents frequently seeded the Greenpeace report as ammunition to fuel opposition to Waste Management projects across North America, which hurt the cause in France.

The two major French waste companies, which were primarily water management companies, resisted the Americans' hopeful expansion. The development team tried, but few prospects responded to their inquiries. The Waste Management team did, however, succeed in buying a landfill company, but winning the facilities' expansions proved elusive, and they were continually stymied by competitors. They added a collection company, Derichbourg, which operated north of Paris. That was it.

Sweden was next on their list. Waste Management was interested in Sellbergs, a subsidiary of a packaging company that was among Europe's largest solid waste operations. Sellbergs had five hundred vehicles to collect and transfer the waste of two million Swedes and four landfills and recycling plants. It had developed an advanced processing system for recyclables called BRINI, which produced a pellet sold as fuel. The company also owned two Spanish affiliates, Ingenieria Urbana S.A. and Saneamientos Sellbergs S.A., which served forty towns in Murcia, Madrid, and Pontevedra.

For Falkman, who negotiated the deal, it was a personal homecoming. His father was born and raised on a rustic twelve-acre farm in Sweden. One of six brothers, his father had been educated in a one-room schoolhouse. Unable to afford leather shoes, he wore thick wool socks in winter's bitter cold. His father emigrated to America in the middle of the Depression, and Falkman still had family there whom he visited. In a full-circle moment, he bought the Swedish company, one of the best and biggest in Europe.

Meetings in search of acquisition candidates taught the Americans a lesson in European differences. On one occasion, they had back-to-back meetings, the first a day in France and the second in the Netherlands. The French served up a five-course meal with fish and meat and a taste of wine in an executive dining room. The next day, the Dutch hosted lunch. They offered the Americans a plate of the thinnest white bread and a slice of cheese, washed down with a glass of buttermilk. "The Dutch were frugal," Falkman learned. The Americans were learning to navigate cultural differences.

As of the early 90s, Waste Management now operated in six European countries: Denmark, Italy, Germany, the Netherlands, Spain, and Sweden. Denmark had added 180,000 customers, and they would later establish a presence in Finland.

The Waste Management executives were eager for more acquisitions. The United Kingdom was next.

The British campaign began in early 1992 when they bought 20 percent of the publicly listed water company Wessex Water plc and formed a partnership with it. "Wessex had a good name," Falkman said. "We felt that gave us local credibility with a solid, well-known utility in the market because a lot of the business was going to be with local councils."

The water company's pricing was regulated, and its executives wanted to invest in businesses outside the reach of utility regulations, in effect using regulated profit to put cash into unregulated businesses.

Together, Waste Management and Wessex bought Britain's Waste Management Ltd, an independent company with a similar name. Weinert and Falkman had known the company's chief manager, Keith Bury, for more than a decade. Bury was a prominent figure and well respected in the British Isles' waste industry. He was also a leader in influencing the European community's emerging environmental rules. The deal and relationship made sense. Wessex and Waste Management then added Wimpey Waste Ltd, owned by Britain's Wimpey construction company, and adopted the name UK Waste for their operations.

THE GLOBAL PUBLIC OFFERING

The London team by then had gathered valuable assets. They would seek more, and the means to do it.

Not surprisingly, Don Flynn had a plan. By 1989, Flynn had become a company elder, a senior vice president, then retired, but Buntrock needed him to take Waste Management International public. He would remain on the Waste Management Board. He had influence. His direction carried weight, and he knew how to use it.

Buntrock, Flynn, and Waste Management International's leaders were planning to sell part of the company to the public. Indeed, Buntrock had started planning for a global offering soon after the Riyadh mobilization in Saudi Arabia ten years earlier.

Buntrock was encouraging Joe Holsten, who had impressed the company's senior leaders overseeing financial affairs in the company's East US region, to consider relocating to London to help manage this global offering as head of finance. Jerry Girsch, an Arthur Andersen alumnus and controller for North American operations who had already shipped several of his team members over, made the call to Holsten. Buntrock invited Holsten to take his wife across the pond for a look-see. It was during the holidays and though it was a promising trip, Holsten worried his family might object. They had made four or five moves already, too often, it seemed, at Christmastime. His oldest daughter spoke up. "I say let's go right now!" she said. The lure of London had won.

Holsten arrived in London as chief financial officer in 1989, another reward for his labors, another step up the ladder. Having a strong financial chief was important. Having experienced operating people was, too, and Waste Management began bringing talent and experience over to prepare for the public offering.

"Don [Flynn] was pretty much of the view, 'Joe, if there's anything you need people-wise, if there's any resistance anywhere, you let me know and I'll talk to Dean. Anything you want you will get,'" Holsten recalled.

They began to interview controllers and managers to insert into countries with Waste Management holdings. Ron Broglio, a Wheelabrator Technologies' waste-to-energy manager, joined the London team to help find opportunities to add more plants to the portfolio. Judy Rosenbeck, a talented engineer, and her husband also came over. Wheelabrator's John Goody, Bill Keightley, and Jim Wegner arrived in search of water and engineering businesses for the company's Rust International subsidiary. Veteran garbage executives came, including Mike Collier as chief operating officer. Bob Coyle and Rich Lauck also provided operations oversight in Northern and Southern Europe.

They adopted the North American financial review process, the disciplined MBRs (monthly business reviews) to start measuring country performance. During one difficult MBR meeting in Italy, Holsten got into an argument with Coyle, a local manager, and learned that there had been a failure to communicate about the timing and strategy of the global offering. "Did you talk to Phil [Rooney] and Dean [Buntrock] before you came over here?" Coyle asked. "Yes," Holsten replied. "They said it's very clear. We want the European business ready to go public as soon as possible."

"They told us we were not to get involved in the operations of the business," Coyle said. "We were to be the overseers, to make suggestions. That was about as far as we were told we should go operationally."

Holsten immediately checked with Flynn. "He said you should listen to what I told you," Holsten recalled. "The company wants to take Europe public, and they want to do it as fast as possible. Nobody expects you to do it this year or next year, but the expectation is that we're going to be selling some stock over here."

RACING AGAINST TIME
As 1992 arrived, Waste Management International continued its

growth. The senior executives added, though cautiously, more staff in London. They moved from offices on fashionable St. James Place near Buckingham Palace to a larger space in a modern building at 3 Shortlands, Hammersmith.

By 1991, Jim Koenig had succeeded his mentor, Flynn, as Waste Management, Inc.'s, chief financial officer. Herb Getz soon became general counsel, a rising attorney with transactional experience. Both were veterans in major financial transactions and had helped take parts of the company public before. This was their international encore.

At Buntrock's and Flynn's direction, Koenig and Getz spent more time in London preparing to launch the initial public offering of Waste Management International common stock. Together, they coordinated the accounting and legal teams. It would be a global public offering with simultaneous listings on both the New York and London stock exchanges. Merrill Lynch, long associated with the company's financings, would be the lead underwriter.

Flynn set a deadline. He was insistent that the offering be no later than April 1992. The London advisers balked. "I had heard over and over from our London advisers that this would not be remotely possible," Getz remembered. "There was a queue that had to be followed."

Flynn wasn't impressed. Nor deterred. He ordered the entire team to assemble on short notice right after New Year's 1992, said Getz, who, sick in bed over the holiday, had been taking his calls. The IPO team, now a large group, gathered in the board room of the Merrill Lynch London office. Rupert Beaumont, a senior partner at the legendary London law firm Slaughter & May, was direct. They could not possibly meet Flynn's April date. But Flynn was steadfast: His date would not be delayed, and anyone uncommitted to it could leave. Lawrence Glascott, a financial executive, and controller Tom Hau, a blunt-spoken Arthur Andersen alum from the South Side of Chicago, helped strengthen the team's efforts. They labored full-time for several weeks and were able to meet Flynn's goal.

In April 1992, Waste Management International sold the public 75 million shares, representing only 20 percent of the international company's shares. The parent and its hazardous waste and waste-to-energy subsidiaries retained the rest. The IPO raised $800 million. Waste Management's leaders used the offering's proceeds to expand, pursuing a mission to be the global leader in environmental services. In its annual letter to stockholders, the company's senior leadership said, "It was among the most successful offerings of its kind ever undertaken." In fact, it was the largest IPO until that time of non-state-owned assets. The offering valued the international company at $4 billion. And Flynn's timing was fortuitous: In May, a financial crisis affected stock markets negatively, preventing any other IPOs for another twelve months.

Following the IPO, the international team raced to strengthen its management team and keep pace with the company's growth. Falkman hired Nigel Wilson as CFO of Waste Management International. Wilson promptly obtained a NatWest Bank credit facility for $800 million and brought in young financial support staff, including Ronan Dunne, who would later become executive vice president and CEO of Verizon Consumer Group, the largest division of Verizon Communications. Wilson would later become the CEO of Legal and General plc, the UK's largest financial institution, with more than $1 trillion in assets under management. In 2022, he was knighted by the queen for achievement in finance and regional development.

The global offering also led to Falkman persuading Bo Gabrielson to join the London team in February 1992 as its chief legal officer. Gabrielson had caught Falkman's attention when Gabrielson served as the general counsel for the seller of Sellbergs S.A. during the transaction's complex negotiations. Gabrielson would later become the international subsidiary's CFO, and in 1997, its CEO.

EXPANSION IN THE FAR EAST

Waste Management International was now in eighteen countries overseas and growing. The company looked toward Asia to find other opportunities.

They developed solid and hazardous waste facilities in Hong Kong and a joint venture in Indonesia. Schroeder found assistance in the American law firm Baker McKenzie, which was based in Chicago, to introduce him to government officials. "I remember in Taipei talking to the deputy minister," Schroeder recalled. "He said, 'You know we're going to industrialize first, and then we're going to worry about the environment.' That made no sense. Nobody would enforce the rules. It was a nonstarter."

Schroeder also recalled traveling to New Delhi to attend a regional meeting of the World Economic Forum in which the company participated. Visiting the deputy minister of the environment there, he found no interest where people relied on waste for subsistence. He found similar circumstances in the Philippines where some people scratched out their existence living atop garbage on landfills. Finding value meant finding rules and finding enforcement.

Hong Kong's industries generated hazardous waste that government officials were eager to address, though. Waste Management International, through its Pacific Waste Management unit, formed Enviropace Ltd., a majority-owned consortium, to develop a treatment facility. Its partners were China International Trust & Investment Corporation Hong Kong Ltd., an affiliate of the People's Republic, and Kin Ching Besser Ltd. The facility launched in 1994. It was one of the world's most sophisticated such operations.

They had drawn experience from the ranks to construct the facility. The team included Mike Rogan, who had been finance director in Riyadh; Bill Grube, an engineer from the company's Rust International subsidiary; Sinon Galvin, from Oak Brook, who came on to lead operations; and Siuwang Chu, from Waste Management's Midwest group. Tom Smith had joined the team. His job, like Schroeder's in Europe, was to find more opportunities. He did so with relish.

Soon after, the South East New Territories (SENT) Landfill opened. It required 150 acres to be reclaimed from the waters of Hong Kong's Junk Bay, adding to the site's unique engineering challenges. A Hong Kong transfer station would be added. In Indonesia, a Waste

Management partnership with P.T. Bimantara opened the nation's first hazardous waste treatment site.

While Waste Management's international expansion moved rapidly across the globe, the company's leadership continued its focus on North America. With increased regulation, managing hazardous waste offered yet another opportunity—and challenges, too.

TEAM MEMBERS LISTED IN THIS CHAPTER

Walt Barber	Bud Ingalls
Larry Beck	Ken Johnson
Jodie Bernstein	Bob Keleher
Ray Bock	Gordon Kenna
Dean Buntrock	Joe Knott
Mike Cherniak	Jim Koenig
Mike Cole	Mike McKinney
Jim DeBoer	August Ochabauer
Jerry Dempsey	Pat Payne
Don Flynn	Don Price
Herb Getz	Matt Radek
Jerry Girsch	Phil Rooney
Milo Harrison	Walt Studebaker
Wayne Huizenga	Peter Vardy

17

WASTE MANAGEMENT RESPONDS TO
HAZARDOUS WASTE

Since the early 1970s, Waste Management's leaders had been eyeing the hazardous waste market and watching where regulations might bring opportunity.

Change was coming, fueled by a legitimate public concern as well as press coverage that painted a portrait of an unregulated contaminated environment seemingly out of control. Love Canal, the notorious, abandoned contaminated site in New York, had exploded in the headlines in August 1978, leaking 22,000 tons of harmful chemicals into surrounding neighborhoods and raising fears among the public of cancer-causing pollution. The public health threat paved the way for the Comprehensive Environmental Response, Compensation and Liability Act (CERCLA), or Superfund, enacted on December 11, 1980, which called for new hazardous waste handling and disposal procedures and for the cleanup of the many abandoned sites that threatened the nation's public health. And there were a lot of them—nearly 50,000 by one EPA estimate in 1970. Pliant politicians were bound to respond. Voters were voters after all. With Superfund, public policy was now taking aim at hazardous wastes. New regulations and laws would soon call for hazardous waste handling and disposal procedures and for the cleanup of the many sites that threatened the nation's public health.

The company's third annual shareholder report in 1973 conveyed the future possibilities of the hazardous waste market: "Recent legislation requiring a phased commitment to proper treatment and disposal of all industrial chemical wastes also represents an enormous opportunity for the company." Waste Management put its money into it.

The leaders of Waste Management then were almost entirely garbagemen focused on the collection and disposal of waste. But a good opportunity was a good opportunity, and the hazardous waste area would be good for business, garbage or not. Waste was waste. Buntrock had planned to build a company that would manage all varieties of wastes, and hazardous waste fit right in, though he would later say he should have been more skeptical of the government's published estimates of waste volumes.

The company's first hazardous waste site, the CID landfill on Chicago's South Side, had served Chicago industry for years, particularly the big steel mills along Lake Michigan's southern shores. During its waste treatment process, the muscular plants produced liquid chemical wastes and acids that needed to be handled before being placed in the landfill. In 1972, the company acquired Chem-Trol Pollution Services Inc., a Buffalo-based company with patents for the design and construction of liquid waste treatment facilities. Chem-Trol added capability to what CID—and soon other sites—would need.

By 1974, CID began sending its industrial wastes to a prototype liquid processing facility. The company formed a technology group to design a larger facility there and lay plans for similar regional capabilities across the country. It also invested in a 3,750-square-foot analytical lab at CID to support its plans to manage hazardous wastes in other locations. In 1975, CID installed an acid neutralization facility enabling it to handle acid wastes from the steelmaking process.

Managing hazardous wastes was a more complex, complicated, and scientific business than handling garbage. It called for lab technicians and specialists who would receive chemical waste samples from prospective customers and analyze them to determine what they were, how they should be treated, and how they should be disposed of. Lots of paperwork was required.

In the fall of 1976, while hunting for acquisitions, Don Price and his associate Mike Cherniak attended a Dallas meeting hosted by the National Solid Wastes Management Association that covered federal hazardous waste regulation. They returned to the company's Oak Brook headquarters excited by the enormity of what they had heard.

Price dashed off a memo or two to Jim DeBoer, the Midwest regional manager to whom he then reported. "The meeting had a direct impact on my personal outlook as to the scope of the business opportunities for Waste Management," Price wrote. At the time, the company's early hazardous waste operations, primarily collection, were handled by its local garbage hauling operations. Price knew there was a bigger market for Waste Management to invest in, and he wanted executives' buy-in. DeBoer, a seasoned garbage executive, expressed little interest in hazardous waste but encouraged Price to take the idea up the chain to Wayne Huizenga. Huizenga got Price's memo. He was interested. At the time, many company leaders were preoccupied with the company's mobilization in Riyadh, Saudi Arabia. Huizenga was on duty, and he told Price to go for it.

The company's leaders knew that only a small percentage of industry's hazardous wastes ever left the generator sites. But for the little that did, they were willing to bet that an investment in hazardous waste would pay off. And so they began to invest.

THE EPA LIST

The EPA helped point the way. Legislation passed in 1970 required the EPA to send the US president and Congress a report on its investigation of the nation's hazardous waste situation. The report, issued in 1974, covered the public health, technological, and economic concerns of hazardous waste handling and included sections on regulation and recommendations. Its chief author was Arsen J. Darnay, deputy assistant administrator for solid waste management.

The report also noted the size of the challenge. The nation was generating 10 million tons of nonradioactive hazardous waste a year: 60 percent was organic, or biodegradable, and 40 percent was non-organic, or material such as glass, aluminum, or plastic that does not easily decompose. Ninety percent was liquid or semiliquid. The hazardous waste volumes were growing annually at 5 to 10 percent. The authors called the current legislative environment "permissive," writing that there was "little economic incentive (e.g., the high costs of adequate management compared with the costs of current practice) for generators to dispose of wastes in adequate ways." The comprehensive report clearly spelled out the problem. Anyone reading could not have missed the message.

The report's end included a section on "findings and recommendations." And there it was: Appendix F: Summary of the Hazardous Waste National Disposal Site Concept." The summary included a list of seventy-four counties that ranked as the nation's best for locating hazardous waste treatment and disposal sites. The EPA had surveyed all 3,050 counties in "the conterminous United States," and narrowed down a final list of suitable sites for handling hazardous waste, emphasizing their health, safety, and environmental attributes. At the very top of the list was Sumter County, Alabama. An asterisk next to the name noted: "Potential site for large-scale processing facility." It was a road map as good as any GPS.

Price and another new associate, Ray Bock, were in hot pursuit. Bock had recently joined the company after Waste Management acquired Bynal Products, which made steel waste containers, and where he had been sales manager. Phil Rooney had assigned Bock to work on hazardous waste sales at CID, which became the birthplace of Chemical Waste Management, a subsidiary focused solely on hazardous waste. Bock would be its lead salesman.

1977: A TOWN IN SUMTER, ALABAMA

In 1977, Price and Bock were on the road searching for locations for hazardous waste facilities nationwide. They knocked on the doors of state and local environmental officials. "I started contacting various states to try to get a feeling if they had any existing facilities that would be capable of handling hazardous waste or could transition to it," Bock recalled. They sought sites across the South and West that had the right geology, a requirement to protect groundwater from contamination.

They had no luck until they landed in the Montgomery, Alabama, offices of Alfred S. Chipley, Division of Solid Waste and Vector Control. Chipley was tall, thin, welcoming, and concerned about the well-being of his fellow Alabamans.

Chipley was just who they were looking for. He told them about an investor group led by Jim Parsons, whose wife, Bobbie Jo, was the daughter of George Wallace, the state's forty-fifth governor and three-time presidential candidate. Price and Bock quickly found their way to Parsons and the other investors, including a man named Mark Gregory. Bock had even attended a meal with Wallace. By coincidence, Chipley had just issued a hazardous waste permit to the investor group for a three-hundred-acre piece of land near the small town of Emelle in west-central Alabama, close to the Mississippi state line.

Emelle was right in the heart of Sumter County, the EPA's ranking for the best location in the nation for hazardous waste disposal. The area lay atop the geologic zone known as the Selma Chalk Formation, a natural chalk barrier seven hundred feet deep dating back millions of years to the age of the dinosaurs. Selma Chalk provided a natural barrier to separate the landfill from the nearest groundwater below.

On Chipley's desk—likely meant to be seen, Price and Bock thought—was a BFI business card. Once again, their chief competitor had gotten there first. The situation called for speed and decisiveness. Price and Bock met with Waste Management's senior executives. "We're saying this is a facility we need to have, gentlemen," Bock recalls telling them.

"'Bockster,'" Bock recalled Huizenga saying, "'are you telling me that we're going to have people from all over the country drive trucks down to Emelle, Alabama, to bring their wastes there?'" Bock said, "Wayne, you're going to have to put up a traffic light because there's going to be a line there."

The investors wanted to sell the property outright. They weren't asking for a "huge number," Bock recalled. Flynn, then CFO of Waste Management, suggested they offer a lower bid but sweeten the price with a percentage of revenue. They successfully negotiated a deal that included a royalty for the former owners, and in 1978, Waste Management had its first acquisition in the hazardous waste business.

At the time, Emelle and its surrounding area were sparsely populated, economically depressed, and deprived of social services. Its residents were mostly Black and poor. The Emelle operation attracted its share of critics, who worried that the tons of hazardous waste trucked in could adversely affect the population's health and safety. Waste Management mounted a community relations campaign led by Gordon Kenna to help boost community confidence in the landfill. While some critics remained, and controversy would occur in the future, the Emelle landfill became an economic development engine for the area. Over time, the landfill's operations served as the area's largest employer and fed millions of dollars of revenue into the state of Alabama and Sumter County, helping improve its schools, build a fire station and town hall, and provide better health care.

Emelle was an ideal site. Years later, in 1985, Kenna, a Waste Management community relations representative, and former EPA official, told *The New York Times* that it would take a drop of water 10,000 years to work its way through the formation to the water below.

1978: CHASING CALIFORNIA

With a site secured in Emelle, Price and Bock eyed California next. Specifically, Kettleman Hills, a low mountain range close to Interstate 5 in the San Joaquin Valley, about halfway between Los Angeles and San Francisco. Like Emelle, its geology and remote location made it ideal for a hazardous waste landfill.

The Kettleman area, reportedly named (although misspelled) after Dave Kettelman, a sheep farmer and cattleman in the 1860s, had experienced an oil boom in the late 1920s in its North Dome oil field.

This was the oil patch. And it was this oil services business, not drilling, however, that brought the Waste Management deal team there in 1978.

Bill McKay operated a fleet of trucks that spread waste oil over the area to control dust. When McKay sought a permit for his oil-spreading business, the state of California issued him a permit to handle hazardous waste in 1978. McKay began operating a hazardous waste treatment and disposal facility, attracting the attention of Waste Management. The company's chief technology officer, Peter Vardy, and his engineers performed their inspections and site suitability checks. Don Flynn approved the deal terms. By 1979, the company had its third hazardous waste landfill.

1979: BURNING WASTE AT SEA

By 1979, Waste Management was operating Chemical Waste Management, Inc., known now as Chem Waste, as a separate division to expand and operate in the growing hazardous waste sector.

The market was big. In 1979, the EPA banned PCBs (polychlorinated biphenyls) and toxic substances found in capacitors and transformers from disposal in landfills. The agency estimated there were 750 million pounds of PCBs in use and several hundred million pounds more in landfills and elsewhere in the environment. PCBs are harmful materials, shown to cause birth defects and cancer in lab animals and a probable cause of cancer in humans. PCBs needed more aggressive management and regulation.

The expectations for growth were plentiful. To address this market, in September 1980, Huizenga jetted to Germany to acquire Ocean Combustion Services, the operator of a 334-foot ship, the *M. T. Vulcanus*, which incinerated hazardous wastes in the North Sea. It was a technology that was proven to work well in Europe and appeared welcome by regulators in the United States. America's need was clear, and with the EPA promoting PCB incineration, Chem Waste had every expectation that approvals would be secured to start US operations. The company planned to bring incineration technology to the Gulf of Mexico. The EPA supported its plan. "EPA considers incineration at sea to be a safe and reliable method of disposing of PCBs," EPA Administrator Anne Gorsuch said at the time, according to the January 1981 *EPA Journal.*

Chem Waste invested in *Vulcanus I* and later built a second ship, *Vulcanus II*, delivered in 1982. Incineration at sea seemed like a good bet.

In 1981, the company sought to destroy 700,000 gallons of PCBs under an EPA-supervised test burn in the Gulf of Mexico. At 1,200 degrees centigrade, more than 99.9 percent of the PCBs were destroyed. A second test burn followed in 1982. The company was paid $10 million for the first test and $25 million for the second. In a *Washington Post* article, reporter Gregory Gordon wrote that "burning toxic wastes such as PCBs at sea, if it can be done safely, is considered a potentially crucial step toward solving the nation's hazardous waste problem."

But opposition and controversy arose after the EPA announced it would grant Chem Waste three-year permits to burn toxic waste at sea. The agency was accused of showing favoritism to the company and skirting proper permit procedures. Communities economically dependent on ocean waters condemned the technology for its potential harm to the ocean ecosystem and public health. Finally, in November 1983, the EPA organized one of the first public hearings on the issue in Brownsville, Texas. More than 6,000 people showed up to oppose the EPA's ocean incineration program. The crowd included coastal residents, migrant workers, church groups, students, and state and local public officials. They gathered support from Greenpeace, the Cousteau Society, and the Oceanic Society. The EPA and Chem Waste faced a lot of heat. Chem Waste eventually gained research permits for the ships, but its effort to gain operating permits stalled. And while the technology had been employed to destroy Agent Orange and the harmful insecticide DDT for government and liquid chemical wastes in European waters for more than a decade, the incinerating ships would never sail in regular commercial service in US waters.

1982: A SHIFT TO LAND INCINERATION

Other acquisitions followed, and Chem Waste now turned to land incineration, putting in place a regional network of treatment and disposal sites across the country. Following the solid waste business model, the company installed regional managers, controllers, and other staff.

Chem Waste added landfills in Port Arthur, Texas, and Lake Charles, Louisiana. It acquired an Ohio company that reclaimed and recycled industrial solvents. In 1982, Chem Waste acquired the publicly traded Chem-Nuclear Systems, a highly specialized business that gave it entry into low-level radioactive waste management, handling materials that had been exposed to radiation, and the nuclear power industry.

Chem-Nuclear had a number of beneficial assets: It had one of the nation's only two operating disposal sites for low-level radioactive waste in Barnwell, South Carolina, and owned a hazardous waste site in remote and arid Arlington, Oregon, about 150 miles east of Portland along the Columbia River. Chem-Nuclear possessed a transportation fleet of forty-five tractors and heavy-duty trailers and owned specialized casks that stored spent fuel licensed by the Nuclear Regulatory Commission. It had mobile units to solidify and decontaminate low-level radioactive waste, and had customers at more than 190 nuclear facilities, government installations, hospitals, commercial laboratories, and facilities generating low-level nuclear wastes. Much of the waste included paper clothing, tools, piping, and metal components that had been exposed to radiation. The nuclear power industry offered opportunity, Chem Waste believed.

By 1983, Chem Waste had 75 plants in operation, 28 more nearly ready for start-up, and 22 scheduled for completion by 1995. Chem-Nuclear's presence offered promise to serve yet another market.

That same year, Chem Waste gained an incinerator facility in Sauget, Illinois, across the Mississippi River from St. Louis. And in 1984, when SCA Services Inc. merged with Waste Management and another company, Chem Waste added hazardous waste landfills in Model City, New York, and Fort Wayne, Indiana, and another incinerator in Chicago permitted to destroy hazardous PCBs. A few years later, Chem Waste would invest heavily in research and development, opening a state-of-the-art laboratory in Geneva, Illinois, hoping to develop technologies to handle its customers' hazardous wastes.

1980S: CLEANING UP ABANDONED WASTE SITES

In addition to acquiring landfills and low-level radioactive waste sites, Chem Waste began cleaning up abandoned waste sites throughout the 1980s. The work became a national priority under the 1980 Superfund law.

The Superfund legislation spread financial liabilities broadly. The liabilities—the cost of cleaning up the sometimes-nightmarish sites—extended to a facility's owners and operators, past owners and operators at the time the hazardous wastes were disposed of there, waste generators and operators who arranged for the disposal or transport of the hazardous materials, and hazardous waste transporters who selected the disposal sites. CERCLA cast a wide net.

To enforce the law, the government would go after deep-pocketed companies to pay for the cleanup costs. They became the potentially responsible parties or, in Superfund parlance, PRPs. And they were some of the biggest, most well-known companies in the nation. The EPA's initial surveys identified an estimated 47,000 such contaminated sites. For Chem Waste, this list was the very definition of a growth market.

In 1980, Chem Waste formed its remedial action unit, ENRAD, to participate in the massive job of cleaning up and restoring hazardous waste disposal sites abandoned by industry and government. (Though Chem Waste soon changed the name to ENRAC when an executive thought ENRAD sounded too radioactive.)

It had competitors, OH Materials, CH2MHill, and Canonie among them. Price, then president of Chem Waste, gathered a group that included managers Mike McKinney and Ken Johnson, Matt Radek in operations, and August Ochabauer, a chemist. Ray Bock became its lead salesman. He had already been calling on executives of major corporations who were beginning to recognize their companies' hazardous waste liabilities.

ENRAC would undertake hundreds of cleanup projects, the first near Gary, Indiana. But few were larger or more remarkable than the 1982 effort in Seymour, Indiana, about sixty or so miles south of Indianapolis and hometown to rocker John Mellencamp. It was ENRAC's first big cleanup.

CLEANUP IN SEYMOUR

The fourteen-acre, fenced-in site was located about two miles west of Seymour, adjacent to its municipal airport where Army bomber pilots had trained during World War II. The area had been previously operated by the Seymour Recycling Corporation (Sears had produced tool handles there earlier), which reclaimed waste chemicals there before it was closed down by the state of Indiana in 1978. It was a massive undertaking: There were 50,000 waste drums (some accounts say closer to 60,000), 98 storage tanks, and several buildings. The drums were stored in single to triple tiers running the length of the site. ENRAC used aerial photography to capture the scale of it. Many of the drums were rusted, damaged, and leaking. The EPA stepped in in late 1980.

IBM had urged Chem Waste to take the project on after dealing with an outfit that had failed to properly dispose of the hazardous wastes generated in manufacturing computers. IBM was not the only company whose waste had ended up in Seymour, but they were insistent that it be cleaned up. And the company knew and trusted Bock.

"As far as taking a look at the environment and their impact on it, there wasn't a company in my experience that was better than IBM," Bock said. He met IBM officials at the site. Bock asked them if they could identify which waste drums—out of more than 50,000—were the company's. He said he was told, "We don't care. We want it taken care of right now. We want that place to get cleaned up, and we will pay the bill." Waste Management was itself a PRP after having acquired a hauling business that had taken waste there.

ENRAC recruited internally and externally to mobilize a crew that understood how to deal with hazardous wastes. A couple of dozen technicians worked on the site. The cleanup process was hard: Workers had to pass physical exams before starting the job. The crews dressed head-to-toe in protective equipment. They wore hazmat suits, called "moon suits," with all openings sealed with duct tape, one manager said. They also wore rubber boots and gloves, safety goggles, and hard hats.

Respirators protected the workers' lungs. Anyone visiting the site had to be tested for properly wearing a respirator. Those who refused were barred. Men had to be clean-shaven, lest the respirators improperly seal against their faces. Cleanup crews sweated through hot summer work-days. They used special trailers equipped with showers to remove any contamination they might have collected. Forklifts moved the drums, some of which disintegrated upon handling. The project included an on-site lab to identify and group the hazardous materials. Some samples were shipped off-site for greater analysis to determine the best means of their disposal. Air monitors were on guard along the perimeter fence, and an alarm signaled if contaminants were identified above threshold limits.

The cleanup, which began in December 1982 and lasted until January 1984, was carefully organized. The drums were grouped into categories: flammable, acidic, water-reactive, and air-reactive. They were treated before being transported to disposal sites. Sludges would be thickened and placed in twenty-cubic-yard trailers lined with plastic to facilitate their handling at the disposal sites and prevent any leakage during trans-portation. A device to empty and crush drums was installed. PCBs were shipped to an Arkansas incinerator to be destroyed. Much of the solid and hazardous wastes were shipped to Emelle, and less hazardous materials to the company site in Vickery, Ohio. Much of the material was loaded onto flatbeds and transported to other permitted disposal sites.

A 1983 *New York Times* story reported that cyanide, PCBs, and arsenic were among the hazardous constituents catalogued on-site. In its story, *The Times* noted, "Farmers reported stunted corn and dead soybeans. Some steers keeled over dead, and one farmer complained that his cattle was coughing."

The crew, as well as staff from the EPA and PRPs, lodged in a local Holiday Inn, where a favorable room rate had been negotiated. The Inn's owners were more than compensated by the steady business at the restaurant and bar, a crew member noted. Eventually, twenty-four companies contributed to the cleanup.

CLEANUP IN ZIONSVILLE

ENRAC teams were busy on dozens of other cleanup sites in the early 1980s while they completed the big Seymour project. ENRAC was also expanding its team. Ken Johnson was assigned to run operations in the East, and Bud Ingalls in the West. Bob Keleher tended to the details of contract administration supported by managers in each region. Walt Studebaker took over engineering and technical services.

In 1983, ENRAC took on an EPA emergency contract in Zionsville, Indiana, just north of Indianapolis. The EPA needed 23,000 drums of wastes, approximately 300,000 gallons of liquids in 53 bulk tanks, and 15 million gallons of ponded liquids removed. The Zionsville Superfund project involved two adjacent sites: the Northside Sanitary Landfill and the Environmental Conservation and Chemical Corporation (Enviro-Chem) facilities, the latter of which was where the ENRAC remediation crews worked to recover, reclaim, and broker solvent oils.

"I remember walking through the Zionsville site in the early morning hours," recalled Keleher, recently back from Riyadh. "The fifty-five-gallon drums at the smaller-than-Seymour site were stacked four to five high. The aisles were like canyons. As the sun heated the drums and their contents, they would expand. It started to sound like a calypso band gone wild with drums banging at different tones for an extended period of time. It was a very uncomfortable feeling."

A ditch flowing south between the sites joined a creek that flowed into another creek that eventually emptied into the Eagle Creek Reservoir, which supplied approximately 6 percent of Indianapolis' drinking water. The threat to public health was clear and unmistakable.

For Chem Waste, managing the cleanups and meeting the strict RCRA and CERCLA requirements that the EPA demanded were occasionally a compliance challenge. Winning the Zionsville project had called for certain activities that were spelled out in the bid but were sometimes not practical on-site. For example, Chem Waste needed to pump and vacuum the areas around leaking drums, but they were unable to do so until the barrels were first sampled and analyzed. If material that was water-reactive got incorrectly pumped into a truck with rainwater, the storage tank could explode. A truck carrying materials from Seymour was involved in an accident and there were issues with its load manifest. The EPA and ENRAC teams were pursuing the same goal, but ENRAC's work had to conform precisely to EPA's demanding expectations or noncompliance ensued. Ongoing coordination and communication with EPA were important.

GREENPEACE REPORTS AND COMPLIANCE ISSUES
By the mid-1980s, Chem Waste had become the nation's largest commercial hazardous wastes manager. It possessed a national network of facilities and the expertise and resources its blue-chip customers needed.

It had had a series of executives running the company, including Price early on, followed by Milo Harrison, Joe Knott, Larry Beck, and Don Flynn. Jerry Dempsey, who had been CEO and president of Borg Warner Corp., became president of Chem Waste in 1985. He would be followed by Pat Payne, Jerry Girsch, and Mike Cole. Some leaders in the Waste Management company considered it a stepchild of the main business of managing garbage.

Running the Chem Waste facilities was much more complicated and difficult than garbage operations. Hazardous waste required scientists and lab technicians and special vehicles. Transporting wastes required shipping manifests. Tracking waste on a "cradle to grave" basis required precise documentation. Absolute compliance was required. Imperfection would be painful not only for the company but for the environment, customers, and communities. Each of the facilities added enhanced capability to deal with the various waste streams.

But the company's capabilities and size also made it a target of environmentalists and federal and state authorities. For some reporters, it made a very good story.

Some years later, in 1991, Greenpeace issued a 285-page report blasting Waste Management, Inc., for the company's compliance failures, of which the most problematic environmental matters were in Chem Waste. The report, called "Waste Management, Inc.: An Encyclopedia of Environmental Crimes & Other Misdeeds," was based mainly on a collection of news clips and public disclosures from across the country over many years.

In its introduction, Greenpeace noted that "managing waste is an exceedingly lucrative business." Greenpeace wrote that "the very name of the company—Waste Management—suggests that wastes can be safely controlled after they are created." Then they added, "Unfortunately, history reveals that this is not true. Once they are produced, dangerous wastes cannot be safely managed." Greenpeace's targeting of the company rested, it seemed, on the company's capabilities in handling waste: "As we come to recognize that 'state-of-the-art' waste disposal technologies don't work, even when practiced by the nation's wealthiest waste hauler, we must eventually recognize that the very idea of disposal is premised on a flawed approach to protecting our natural resources and ourselves."

Greenpeace was not alone. A lesser-known group, the Citizen's Clearing House for Hazardous Wastes, Inc., also issued a critical report. Both organizations appeared to be calling for a zero-waste world. For them, Waste Management and Chem Waste were providing solutions to manage wastes and, therefore, forestalling their more utopian aspirations.

The criticism wasn't entirely undeserved. Chem Waste had, indeed, suffered a series of compliance failures at several sites, which were well publicized in *The New York Times*, *The Wall Street Journal*, Fort Lauderdale's *Sun-Sentinel*, and in regional newspapers around the country where interest in company facilities was high. It didn't help that during this time, the environment was in the news, and so, too, would be the company. (The company's explanations, however reasonable and legitimate, sometimes were more complicated and involved than the news coverage offered, and the circumstances more extenuating.)

At the CID site, for example, there was an allegation that the facility had received "DCB"—dichlorobenzidine, a suspected carcinogen—without a permit for several months in early 1980. In fact, the site had such a permit. But it had failed to provide state authorities with required manifests for the material, which was a violation. It also had received certain wastes for a brief period during which it had a permit renewal pending, not an actual permit, another failure.

Chem Waste's Ohio Liquid Disposal site's receipt of certain wastes from Canada and handling of PCBs were other problematic instances. At the Emelle site, the handling of PCBs from cleanup programs and drums containing PCBs, the storage of transformers, record-keeping related to hazardous waste burial locations, and leachate management also raised problems. Chem Waste paid a $3,000 fine for sending Seymour cleanup wastes to two sites that, although permitted to receive the material, were not listed in the project's work plan. There were several other violations and issues.

Its reputation questioned, Chem Waste hired the law firm Karaganis Gail & White, Ltd., to undertake an independent investigation of its early environmental actions and issue a report. The Karaganis investigation noted the bad publicity: "Leader in Toxic Dumps Accused of Illegal Acts"; *The New York Times*, Monday, March 21, 1983; "Illinois Accuses Waste Management of Scheme to Hide Illegal Toxic Shipments"; *The Wall Street Journal*, Tuesday, March 22, 1983.

The Karaganis report also straightforwardly cited the failures. It elaborated on the circumstances of each, including failures to communicate among staff and with environmental agencies, and not meeting technical permit requirements. It highlighted cases of media hyperbole and misunderstanding. For charges that had merit, Karaganis explained they fell into three categories: technical violations that were not significant;

violations caused by reliance on informal variations by regulatory authority; and violations that, while not harmful to the environment, were regulatory lapses. Still, meeting regulations was demanded.

Karaganis also offered an observation on Chem Waste's environmentalism: "Protection of the environment must be the overriding factor in the company's hazardous waste treatment and disposal operations. The central question is not regulatory compliance or public perception—but whether the company's hazardous waste operations protect the environment. Based on our discussion with company technical personnel, the company's technical approach to hazardous waste management is highly consistent with sound environmental policies."

Finally, the Karaganis report offered several recommendations the company was required to follow. The first could not have been clearer: "Vigorous compliance with the letter of the law."

CHEM WASTE RESPONDS

To help address these issues, the company needed some additional heavy-weight talent. Walt Barber, the former acting administrator of the EPA early in the Reagan Administration, joined the company in 1983 as vice president of environmental management, a function like Peter Vardy's in Waste Management's solid waste business. He had also headed the agency's air quality planning office during the Ford and Carter years. Barber, highly respected among his Chem Waste colleagues, had come to the company from Jacobs Engineering. Jodie Bernstein, a former EPA general counsel, joined Chem Waste as its general counsel. Chem Waste was adding regulatory agency experience and understanding to its player bench at an important time. The company needed to re-emphasize its commitment to compliance and create a compliance officer position, internal audit programs, and an environmental management system to ensure that its operations met regulatory requirements.

Chem Waste's management had learned a difficult lesson. Still, lapses and human failure would never entirely escape the company.

By 1986, Chem Waste was fairly mature, its network of assets and capabilities in place. It was looking forward to a profitable future. The senior executives of Waste Management thought so, too. But the business was still growing at a rate much faster than the solid waste business. Don Flynn, with support from Merrill Lynch, saw an opportunity to capitalize on this and provide the company with additional growth capital on very attractive terms. Flynn tasked Chem Waste CFO Jim Koenig and Herb Getz to prepare Chem Waste for an initial public offering.

Koenig, an ascending young executive, had returned to Oak Brook after being Weinert's "wingman" helping to build the Camp in Riyadh. Koenig had even driven trucks for the project's difficult start-up and had spent the previous six months commuting between Oak Brook and Riyadh. Upon his return, Koenig had worked as Phil Rooney's assistant and teamed with Bill Debes in performing acquisition analyses. He had played a lead role in running Chem Waste's site remediation division before becoming its chief financial officer.

The executives saw more growth, more shareholder value in the future from both the parent and subsidiary. The newly public company would continue to expand, they believed, as it added new capabilities to serve more industrial customers who understood the weight and force of ever-tightening federal RCRA and CERCLA rules.

In October of 1986, Chem Waste provided the public with an opportunity to invest in the more specialized hazardous waste business when it conducted an initial public offering of stock. It sold 18.9 million shares domestically and abroad. The proceeds of $308 million went to pay a dividend and reduce its indebtedness to its parent company. The shares, carrying the ticker CHW, were listed on the New York Stock Exchange. Waste Management retained 81.1 percent of the shares.

Chem Waste was growing. But the Waste Management executive team had not ceased its work to find more growth.

TEAM MEMBERS LISTED IN THIS CHAPTER

Mike Andrews	Joe Holsten
Ken Arnold	Bill Hulligan
Amy Burbott	Al Morrow
Don Clark	Dennis Nowatarski
Rich Evenhouse	John Reinecke
Vito Galante	Phil Rooney
Jerry Girsch	Ben Victory
Joe Graziano	Jane Witheridge
Dennis Grimm	Peter Yaffe

18

PROGRESS IN PENNSYLVANIA

In 1987, a barge carrying 3,000 tons of New York garbage captured national attention. The Mobro 4000 garbage barge had been part of a scheme to transport New York's trash to North Carolina. Every day, the city's trash was exported to other states' landfills. The barge spent six months meandering along the eastern seashore traveling as far as Mexico and Belize to find a port that would accept its cargo. It was unsuccessful and finally returned to New York to unload its waste. The headline-grabbing voyage generated fears of a nationwide disposal crisis. The problem was disposal space, or a lack of permitted facilities, but there was not a lack of land or places to handle it. Local governments had put off politically difficult siting approvals, creating the problem. Now, with a crisis looming at their doorsteps, local public officials, particularly in the East, were in a mad rush to find solutions—and available disposal space—to address the concern.

Adding to the pressure were new legislative mandates in New Jersey and Pennsylvania requiring local governments to implement longer-term solid waste plans that designated where their garbage would be disposed of. All of this seemed to come at once.

By then, Waste Management had completed its integration of the SCA

Services assets, many of them in the East and some in Pennsylvania. Its network would be uniquely positioned geographically to respond. Other regions across the country also experienced disposal space shortfalls, but the problem was particularly acute in Pennsylvania, and the company was on a mission to help solve the problem. Waste Management welcomed these challenges and began expanding rapidly in Pennsylvania.

Critical to this expansion was a young manager named Dennis Grimm, who would become the company's senior operations executive in Pennsylvania and New Jersey. He had joined the company during the SCA acquisition, and the geography he knew so well was ground zero in the perceived crisis.

FAMILY FARMER TO GARBAGE EXECUTIVE

Grimm understood the business and the need. His family had had a successful collection company that operated in and around Lancaster, Pennsylvania, that had become part of SCA. Like so many family garbage businesses, the operation began more humbly by collecting waste that was fed to pigs on the Grimm family farm. Over time, SCA recognized Grimm's talents, and he became the regional manager for Pennsylvania and five surrounding states and then its vice president for the broader East Coast. He knew the territory and built up SCA's presence by buying companies, doing tuck-ins, and after certain SCA officials were implicated in federal and congressional investigations for alleged organized crime ties in the early 1980s, even divesting a group of New Jersey businesses, a number to their former owners.

"We were the biggest in Pennsylvania," he recalled of his SCA days. Grimm said he would often run up against BFI but faced little competition from Waste Management in the state. In doing deals, his reputation preceded him. He grew the former owners' positions. And he worked well with state regulators. His SCA approach resembled Waste Management's.

In the days leading up to SCA's merger with Waste Management, in 1984, Grimm got to know Bill Hulligan. Hulligan had also been an SCA employee. While Grimm was not directly involved in the negotiations, he became friendly with Hulligan during the process. "He was wining and dining us," said Grimm. "Bill and I went to Atlantic City." Grimm said SCA wanted Waste Management to win the merger. Waste Management sealed the deal when its cash offer trumped BFI's stock proposal.

When Grimm arrived at Waste Management, he began reporting to Rich Evenhouse, who was then in charge of operations in the Northeast. Evenhouse was not particularly enamored with SCA people. Yet the two grew to trust and like each other. Grimm was responsible for operations in Pennsylvania, West Virginia, Maryland, New Jersey, and parts of New York, then a "relatively small region," he said.

Waste Management had a lot of capital. "So, I went out and bought what I could buy that made sense," Grimm said. Landfills were of particular interest. He had good rapport with Pennsylvania state environmental officials who knew he had a record of addressing any environmental problems that surfaced at sites he had overseen while at SCA.

On one occasion early on, Mike Andrews, a Waste Management engineer, was trying to arrange a meeting with Nick DeBenedictis, the director of the Pennsylvania environmental department, through a consultant. Andrews interrupted a meeting between Evenhouse and Grimm to share his excitement at the possibility of achieving such a sit-down. Grimm, who had not been consulted and knew DeBenedictis well, immediately picked up the phone and got DeBenedictis on the line. "Hey, Nick. This is Dennis. My new boss is here from Waste Management, and I'd like you to meet with these guys. Is it okay if I bring 'em in?" A meeting was set for the next day. "I was sure they didn't think we had those contacts, but we had as many or more than they had," Grimm said. It paid to have friends. Especially in important places.

WASTE MANAGEMENT GROWS IN PENNSYLVANIA

The company was growing rapidly in states across the country, but its performance in Pennsylvania would be nothing less than outstanding, thanks to its network of landfills and astute disposal agreements with local governments, including the city of Philadelphia. The sites included

Waste Management's GROWS (Geological Recovery Operations and Waste Systems) facility north of Philadelphia along the Delaware River and the Lakeview Landfill up in Erie; the former SCA Pottstown Landfill in Chester County and Modern Landfill in York County; and the Tullytown Landfill, a new "greenfield" facility just to the south of GROWS. The Tullytown facility moved from a zoning change to accepting its first load of trash in 1988 in just eighteen months, an incredibly abbreviated period for a project that normally took years. It became one of the state's largest disposal sites.

By 1987, there had been talk of closing GROWS, but the team gave it new life. York County's Modern Landfill, which had environmental issues, was a special situation. "I convinced Waste Management to put it in the group," Grimm said. "I told them, 'We have to have this landfill. We will get it expanded.' No one ever got a Superfund site expanded—ever. That was the first Superfund-listed site in Pennsylvania to get an expansion, not only an expansion but it is still operated today."

To get the expansion permit, the company responded with an expensive remedial program at the Modern facility that satisfied the regulators. "We looked like heroes to the state. It satisfied a huge need for disposal capability for the state and municipalities," Grimm said. "Everybody at Waste thought I was crazy. 'You're gonna get a Superfund site expanded?' Yes, we're gonna get it done." They did. All the Pennsylvania sites would ultimately gain expansions.

Grimm had a good record with regulators. His environmental management lieutenant, Vito Galante, had come from the steel industry, where disputes with regulators were more commonplace. Grimm encouraged him to accommodate the officials: "We're not going to fight them. They were our bosses. He got that out of his head," Grimm said.

Waste Management's landfill and transfer station activity coincided with worried discussions among local officials in New Jersey and Pennsylvania about where to dispose of their citizens' trash. The two states required counties to have solid waste plans that ensured they had places to dispose of their waste. State officials were also concerned about laws prohibiting moving waste across state lines even though interstate commerce in garbage was common elsewhere across the country. Waste Management addressed these challenges and concentrated on expanding its landfills and their capacity.

They secured New Jersey transfer stations in Essex County and Avenue A in Newark, the latter with an emergency order from the governor. And they would secure two more in Philadelphia: The Forge and Philadelphia Transfer sites. They even won a contract to operate the New Jersey Meadowlands balefill disposal site. Considered a special project, the site's waste was cubed into bales by six antiquated machines and stacked like hay in the site.

At one point, the company's transfer and disposal sites were handling up to 50,000 tons of waste a day. "It was the most profitable group of companies Waste Management had and probably ever had before," said Jane Witheridge, then the district vice president responsible for landfills and transfer stations in Pennsylvania and New Jersey. "It was a $250 million per year group of companies."

THE CORE TEAM

In 1986, Witheridge joined the region's leadership team from the Oak Brook environmental management group. She reported to Grimm. His team included environmental engineers, lawyers Amy Burbott and Ken Arnold, Peter Yaffe in government affairs, and Joe Graziano, an exceptional asset in navigating local government and winning bids. He had joined during the 1980 acquisition of The Warner Company's Waste Resources operations.

Witheridge worked closely with Grimm and Graziano, the latter of whom "knew RFPs better than anybody. He thought about every single angle of every single contract," said Grimm. Graziano's talents stood out, his contributions well known. "What Joe did for us was big time. We would have a big bid coming out for, let's say, the city of Philadelphia, or whatever, a big project," Grimm said. The Waste Management team would request pre-bid meetings and get bid specifications, which might be dozens and dozens of pages.

"Most guys who go to a pre-bid meeting go to listen. He would come to these pre-bid meetings with a presence about himself that exuded confidence, and he got most of the changes put into a lot of these real, real big bids, and in a lot of cases that was to our benefit," said Grimm, whose fondness, respect, and gratitude for Graziano could not be understated. "He had studied those specifications for days. He would tear them apart, take out the stuff that was meaningless and was going to cost the municipalities more and was going to put us at a disadvantage for whatever reason. He would work nights doing these things. I'd get a call from him at eight o'clock and he'd say, 'Dennis, I found this, I found that.' It was just unbelievable."

Joe Holsten also joined the team, a peripatetic accountant who became the region's controller in 1987. But first, his background. Holsten had grown up on a farm in Fairmount, Indiana, in Grant County, about halfway between Fort Wayne and Indianapolis. His was a town of fewer than 3,000 people. He majored in sociology and was in ROTC at Indiana University, and after graduating in 1973, he enlisted in the Army, where he earned a top-secret clearance classification for his work in intelligence. Four years later, he returned to Indiana's Kelley School of Business and graduated first in his MBA class. His specialty was finance, which secured him a job with the Coopers & Lybrand accounting firm in Tucson, Arizona. While there, he earned his CPA and worked with mining interests until he "met a guy in Phoenix working for a start-up garbage company out of Chicago," he said. He was told, "The biggest problem this company has is they are growing so fast they can't bring people on fast enough to accommodate the growth."

He was soon working at Waste Management's Phoenix office, which boasted thirty-five front-end routes and thirty-five roll-off routes and great productivity numbers. "We spent most of our working nights trying to figure out how to get a landfill," he said.

His employment with Waste Management worried his mother. He had called to tell her he was changing jobs. He explained it was a finance position. His mother wasn't sure what his new company did, despite its descriptive name. "She said, 'I still don't understand what you're trying to tell me.' I said, 'You know, the truck that goes up and down the alley twice a week to pick up the garbage.' She started crying. I said, 'Mom, it's a great job, great people.'"

"We thought you were doing so well," he remembered her saying. "She thought I had failed, and it was the only job I could get working for a garbage company. She always hoped she could tell her buddies that her son was a doctor or lawyer." Mom needn't have worried.

After several years in Phoenix, Holsten advised Girsch, controller of the North American solid waste business in Oak Brook, that he was thinking about leaving the company to be the chief accounting officer of a small hazardous waste business in Oakland, California. Girsch told him to catch a plane back to Chicago that night. "I won't try to talk you into making a bad decision," Girsch told Holsten. "I will try to convince you that you're making the wrong decision."

Holsten's consideration of the switch surprised even the hazardous waste company's chief executive. "You work for the best company in this industry," Holsten recalled the executive saying. "'My daughter has a recital tonight that I'm missing, and I'm here talking to you. You're really kind of the last person in the world that I want to be spending the evening with. And among other things, I can't believe you're here. You're a rising star at Waste Management, and there's no reason in the world that you would ever consider working for us.' He kind of convinced me that working for them wasn't such a great idea after all."

Girsch had a bigger role in mind for him. In 1983, Holsten moved to Atlanta as district controller working closely with John Reinecke, the district manager, in the region governed by Al Morrow, who was "quite a character" and "a no-nonsense guy," Holsten recalled.

Three years later, Holsten moved again, this time to New Orleans as assistant district manager in Don Clark's growing southern domain. The district included Louisiana, Mississippi, Alabama, and the Prince George's County Landfill in New Orleans. Holsten wouldn't be there long. The plan was to move him to Atlanta next to succeed Reinecke. Holsten and his family were all packed and headed for Dogwood City when Hulligan's call came. "You can continue to pack so you can move, but we want you to move to Philadelphia." Rooney also called: "You're the first guy we moved before we completed a move," Holsten recalled Rooney saying.

So Holsten arrived at the group's offices near the GROWS landfill. The senior East team was soon forced to move to provide additional staff space for work on planned site expansions, and they worked alongside local district employees in a new office building in nearby Bensalem.

Holsten was not altogether unfamiliar with the eastern landscape. Earlier in his career at Waste Management, he remembers being called to work on the financial due diligence review of the SCA merger. During the review, he and another controller, Dennis Nowatarski, were to meet an SCA manager at the Detroit airport to start the process. The SCA representative suggested they sit down, and the more-to-the-point Waste Management men said, "Why don't we go over to the office and start getting background information?" The man replied that there was no reason to go to the office; he could tell them everything they needed to know at the airport. Hulligan checked in to see how the two were progressing with the evasive SCA representative. Holsten faced an informational roadblock. "He gave me three sheets of paper, and that's all he gave me," he told Hulligan. "I'll give him a call," Hulligan said. The next morning, Holsten and Nowatarski met the man in the conference room of the SCA Detroit office. "What do you want to know?" he told Holsten and Nowatarski. "This room is yours." The SCA man had conceded. He provided them with piles of books and statements and contracts. The books were opened. Hulligan had powers of persuasion.

Holsten arrived as the region's controller in 1987, but his role was more that of a chief financial officer, Grimm would say. Grimm had grown the region "like crazy" and was a "great manager," his colleagues agreed. But with garbage crises, anxious municipalities, new landfills, new site expansions, and new governmental contracts, there was a lot to keep track of. Holsten was there to help Grimm and to keep a close watch over the numbers and any trouble lurking in them. During his tenure, he helped Grimm continue to expand.

"We were nicknamed by Mike Rogan the Landfill Barons," Holsten remembered. "We had just recently opened the GROWS Landfill and Tullytown. We just got a massive expansion at Pottstown, a huge expansion at Modern, and a big expansion at Lakeview. Landfill pricing in our region was skyrocketing. We were getting up to $75 a ton range in 1988 and 1989. The state and some of the cities were going crazy trying to figure out where they were going to put their garbage, Philadelphia being one, and Mercer County, New Jersey, being the other."

MOVING GARBAGE OUT OF PHILLY

Mercer County needed air space, that is, a place where their waste was guaranteed to go for burial that they could include in their solid waste plans. Waste Management had just the space in Pennsylvania. But the Jersey officials had agita about exporting waste into the state. "They were concerned that the feds would come in and tell them they were breaking the law by exporting waste to out-of-state sites," Holsten said.

Graziano, the clever marketing whiz, had a plan. Waste Management would just sell the county air space for $30 million. "A $30 million transaction of this unusual nature was viewed as quite a feat," Holsten remembered. So severe was the county's need—and the economic vagaries of supply and demand to them so worrisome—that disposal prices had risen to nearly $100 a ton in the Philadelphia market. Years later, Holsten would still have a copy of the $30 million check in his office.

Waste was flooding in from New Jersey, and in 1988, Grimm and Holsten negotiated a huge transfer deal—their biggest—with the city of Philadelphia. They signed the contract on June 30th. Still more waste was coming from Montgomery County, Pennsylvania, northwest of Philadelphia. The Waste Management team had to scramble across the company's network to find enough big semis to haul all of it.

The City of Brotherly Love's garbage was planned to be routed to the new Tullytown facility. The Waste Management team was running hard and fast to complete construction of the site and open its doors. Witheridge asked one of the team's lawyers whether he had sneakers with Velcro fasteners. "You are going to have to run so fast and you won't have time to tie them," she deadpanned.

At times there could be intracompany tension between the hauling people and the landfills. The hauling companies were collecting the waste, the transfer stations were filling up, and the pressure was on the landfills to ensure capacity was built and ready. A fatality during Tullytown's construction gave Witheridge pause. She recommended that the work be slowed down, but the crews pressed on. Pennsylvania required hard-to-find gravel for the site's liner system. Engineer Ben Victory found the material in Canada, and it was shipped to the site's Delaware River dock.

The day before Philadelphia's trash was due to arrive at Tullytown's scales, Witheridge called Grimm: "Dennis, the state hasn't been here yet." The state's sign-off was needed before the site could open. It was 1 p.m. Grimm knew who to call. "We have all this trash coming," he told the official. "Your guy hasn't walked through." It would be done, he was assured. Finally, around 3 p.m., the state officials came. Witheridge took them around the site, they signed off, and Tullytown was ready. The next day, trucks bearing Philadelphia's garbage lined up at the gate and were welcomed into Tullytown.

The feat was worth celebrating, and the staff did, with friends, officials, and invited guests. "The event was unlike any other open house I have been part of or attended," Witheridge remembered. The managers wore black tie and evening wear. An orchestra entertained. Hors d'oeuvres and drinks were plentiful. They had earned them. The Pennsylvania team had created a valuable network. And it was being put to good use.

TEAM MEMBERS LISTED IN THIS CHAPTER

Paul Aeschleman	Rosalie Mule
Larry Beck	Gary Peterson
Dean Buntrock	Steve Ragiel
Stu Clark	Phil Rooney
Jerry Girsch	Bob Russell
Bill Hulligan	Serge Sterelli
Tom Kaczmarski	Vince Sterelli
Dan Kemna	Kent Stoddard
Jim Koenig	Tom Tomaszewski
Ryan McKendrick	Don Wallgren
Bill Moore	Chuck White
Frank Moore	Jane Witheridge

19

THE RISE OF THE NATION'S
RECYCLING LEADER

Just as the company celebrated its landfill expansions in Pennsylvania and elsewhere, a different challenge arose: recycling.

"The country was moving to recycling and new technologies were being developed to handle the materials to be recycled," recalled Jim Koenig. "The move to recycling had a significant impact on the company's ability to maintain margins and to sustain profit growth in the US." The company's most senior leadership saw disposal prices rising, which affected its hauling operations profits. Greater consumption of recyclables was occurring, and the public was calling for it.

Waste Management, indeed, the entire hauling industry, was on the front lines of where recyclables could be collected most easily. The practice on a measured commercial basis was not new to Waste Management or its competitors, either. For years earlier, haulers had collected old, corrugated cardboard and paper from their customers and resold it when higher prices made it worth their while. They had recycled metals at scrap yards and paper fiber at paper stock plants. Newsprint makers had sought fiber from "urban forests." Some cities such as Madison, Wisconsin, and Chicago had started to test residential recycling programs in the early 1970s, installing bins under the rails of collection truck chassis for residential newspapers. Transfer stations had balers for cardboard. Storage space was required for bales when fiber prices were low and the activity wasn't worth pursuing.

But now, the company needed to learn how to respond to the increased demand for recycling, manage it, and grow with it. Like other issues the company had faced before, the solution would not come easy. Buntrock was committed to leading the push for recycling, but he also knew it would not make any money. The company's teams had to feel their way. And they weren't quite sure how to take on this new service.

By the mid-1980s, recycling efforts were ramping up. The company was committed to lead in managing North America's waste stream, which would now include recycling. It became a top priority for the company, and Waste Management's team members embraced their role in it. Their efforts paid off, as Waste Management would become the industry recycling leader.

RECYCLING ROOTS IN THE US

The company got off to a modest if determined start. Its first recycling efforts actually focused on waste reduction. In its first formal mention of recycling in 1986, Waste Management told shareholders that it had "launched exciting new programs that we are confident will place us in the forefront of efforts to reduce the volume of wastes requiring disposal and to increase the recycling of usable materials from the waste stream."

In the years prior, citizens across the US had begun to encourage their communities to launch new residential recycling services. There was no one-size-fits-all community recycling system.

The first programs began in the early 1980s, with different regions testing new collection approaches. Across New England and in New Jersey, a dual-stream collection system featured a single household bin for recyclables, and the materials (typically old newspapers and metal, aluminum, and glass containers) were sorted into a truck with two compartments. In Ontario, Canada, residents tossed all their recyclables into a single blue bin, the contents of which a driver fed into a four-compartment collection truck. In Northern California, households employed a novel system that included three different-colored plastic bins, much like stackable milk crates, and the materials were sorted by the driver at the curb into the truck's paper, container, and glass compartments. The systems were residential recycling's first steps, and collection was slow and inefficient.

With experience, the company would learn how to respond and proceed.

THE LITTLE RECYCLING COMPANY THAT COULD

In the early 1980s, an independent company, Empire Waste Management, was pioneering Northern California's first residential recycling collection program in Santa Rosa. The Empire program was a cooperative effort undertaken with the state of California and the city of Santa Rosa. And its success required changing the entire city's recycling behavior.

Thirty-one-year-old Stu Clark managed the recycling operation. He had started as a sorter of recyclables, became a collection driver, and then a manager of Empire's curbside recycling program. Clark knew municipal recycling as well as anyone in the country, and he engaged a public relations expert, Alan Milner, to help change residents' behavioral patterns. Milner had run a successful water conservation campaign for Sonoma County and Santa Rosa, and he helped Clark run one for Santa Rosa too. "Alan is actually the guy who came up with the idea of providing everyone with those stacking containers," Clark recalled. Milner built a mock-up of stackable containers that residents could use to sort and recycle in his garage, and Empire would go on to work with a plastics company to manufacture and secure the rights to sell them. The Santa Rosa program was a huge success, and the city of San Jose, about one hundred miles south, took notice.

In 1985, San Jose asked for bids on a contract to run a curbside recycling program separate from waste collection. Empire competed against industry leaders Waste Management and BFI for the bid. The city had $100,000 it wanted to put into a one-year pilot program but would extend the contract to five years if the winning company succeeded in hitting certain metrics, one of which was achieving 25 percent community participation. Empire easily surpassed the test.

"It was clear that Waste Management and Browning-Ferris really didn't have an understanding of how to make money in recycling or how to run the program," Clark said. "As great as both those companies were, Empire had been doing recycling for decades and doing the curbside version of it for a number of years." The city surprised the industry leaders when it awarded the recycling contract to Empire, which applied its Santa Rosa know-how to San Jose. "It was kind of a wake-up call for Waste Management and the whole industry. Here's a company that could end up with a citywide contract for recycling side by side with garbage collection," Clark said.

Empire, the small independent company, understood that it could leverage its recycling success into a much bigger garbage collection contract in the future.

"At Empire, we saw it as a way for a little company to come in and say we might not be big enough to bid on that garbage contract, but we will be in a few years," Clark said. Empire's plans called for it to take its twenty-thousand-home pilot in San Jose citywide. Just maybe, it might win the waste collection contract, too. Indeed, San Jose officials thought Empire's recycling system was one of the most successful programs it had ever implemented. "The city council person in whose district we put the containers came to me at one point and said, 'Stu, I have to thank you because you've just cemented my reelection for three terms.'" It was unquestionably popular with the public, and San Jose became the catalyst for recycling's rapid expansion elsewhere, Clark said.

In 1985, the same year Empire's pilot began, San Jose awarded Waste Management its waste collection franchise, which reached both households and businesses. Then in 1986, Bill Hulligan, who oversaw the company's North American waste operations, led Waste Management's acquisition of its recycling competitor, Empire Waste Management.

With it came the San Jose recycling program, now the first major city curbside collection program in the country. After a phased-in start and metrics drawn from two communities—Berryessa and Willow Glen—the program eventually reached 186,500 households. It came with a 40 percent pretax profit. Waste Management branded its service Recycle America (and Recycle Canada north of the border), and Clark helped lead its expansion up and down California and across western states as he rose through the company's ranks. The California project provided a prototype for the rest of the company to learn how recycling could be done—even though many managers were not yet embracing it.

Division managers had spent their careers relying on waste collection for profit. Some saw recycling as an unwelcome burden. Some plainly resisted it. Recycling required labor, new vehicles, and processing equipment, and it exposed the company to unpredictable commodity sales often complicated by geographic location. It added a substantial layer to costs and made municipal contract bidding and negotiation more difficult. These managers began to recognize, however, that increasingly, municipal contracts required recycling services. If they wanted the waste contracts, recycling was now going to be an important part of the business.

Clark needed to persuade the doubters in Waste Management's California divisions that they could profit from the enterprise. His irony helped. "Look, guys, I know recycling's a pain," he'd say and then add in his friendly way, "it's just like the garbage business; you can't make any money picking up the trash." Silence and rebuttal would typically follow, Clark said, along the lines of, "What do you mean? We make lots of money picking up the trash." "How the heck do you do it?" Clark would ask. "You gotta pay the drivers, you gotta buy those trucks, do all that maintenance, buy the fuel, pay for insurance, pay for customer service, and then the stuff you're picking up, you take to the landfill, and you pay somebody to dispose of it. I don't see how you can make any money doing all of that." The managers would respond, "Well, we charge a rate." That's when the light bulbs came on, Clark said. The company would charge for the service. It was a turning point. Cities recognized recycling's popularity and began to see it as an essential ser-vice they would willingly pay for. The company started to incorporate profit-generating rates into franchise and other collection contracts calling for recycling.

RECYCLING RAMPS UP

Rooney championed recycling. Hulligan implemented it. Eventually, many regional and district managers got the message and got on board. Rooney and Hulligan understood the necessity of providing recycling services and knew it would not be easy. They also recog-nized that inefficient programs and disadvantageous contracts could damage the company's financial performance. They were intent on better understanding their predicament, and they encouraged divi-sion managers to develop solutions that responded to local situations. In California, companies were typically awarded exclusive franchises in which increased service rates could be more easily adjusted and accepted. Elsewhere, open markets meant tougher competition, where it would be harder. Persuading some customers to pay for recycling and even accept the up-and-down risk of commodity sales would be a challenge. Officials in local governments thought recycling should cut their costs. They did not believe it raised costs.

In 1986, Rooney recruited Bill Moore, an environmental specialist focused on hazardous waste reduction, to be the company's first recy-cling director. A handful of employees in local divisions across North America had also been residential recycling pioneers.

While Moore had specialized in industrial hazardous waste manage-ment, his new responsibilities took him in a different direction. "I virtu-ally got a whole new career when I went from the hazardous industrial chemical industry over to Waste Management," he said. Within a year of his arrival, Moore's focus shifted entirely to solid waste recycling.

In Oak Brook, Moore's newly established corporate recycling team was providing support across the country. Moore recalled Rooney coaching him "'to make twenty decisions every day. Get twelve of them right and don't worry about the other eight because that's a winner. You're batting a winner.'" Moore assembled a group to find answers. It included Tom Kaczmarski, who had a banking and manufacturing background and supported local managers by developing pricing pro formas and offering guidance in developing processing facilities. "I developed a binder for material recovery facilities," Kaczmarski said. "It was a step-by-step, hold-your-hand-kind-of effort to walk the managers through the entire learning process and give them an idea of what they could anticipate."

Over time, the group expanded to include Dan Kemna, Paul Aeschleman, Ryan McKendrick, and others. Among those working on recycling locally were Tom Tomaszewski, Gary Peterson, Vince and Serge Sterelli, and Steve Ragiel, who would go on to be president of Recycle America. The team reported to Don Wallgren, the environmental management vice president.

Meanwhile, recycling volumes were growing. In 1981 the US recycled an estimated two million tons of materials. By 1990, that number grew to 16 million tons. In five more years, it reached 26 million tons, according to the EPA and industry reports.

Legislation skyrocketed. Between 1989 and 1993, nearly 3,000 state recycling laws were proposed, of which about 500 were enacted. The laws fueled recycling volumes. States mandated recycling goals. The most important legislation developed in California, and the Waste Management team there played a critical role in creating new groundbreaking legislation to champion recycling, in partnership with state environmental organizations.

GROUNDBREAKING LEGISLATION IN CALIFORNIA

As a state, California led the way in recycling legislation and reform.

The company's team there included Kent Stoddard, who joined in 1988 to head state government affairs in Sacramento; Bob Russell, a consultant who had served in the Carter White House and had connections with California environmentalists; Chuck White, a politically savvy engineer who handled state regulatory affairs; and Stu Clark, the recycling pioneer from Empire Waste Management. It was a strong team that not only had close ties to environmental leaders but shared in their support for pro-environmental policies. Stoddard had been the chief environmental consultant to California Speaker Willie Brown and earlier worked for the state's House Ways and Means Committee, specializing in hazardous waste issues. He also had worked for Governor Jerry "Pat" Brown. At Waste Management, Stoddard "got to do environmental work on a corporate budget," he said.

During the late 1980s, the company had been working to respond to attacks by some environmental groups. Russell estimated there were about eighty environmental groups in the Bay Area alone, some more hardline than others. Larry Beck had flown to California to try to find a way to respond to the company's environmental critics there. Frank Moore, the company's government affairs chief, suggested that Beck meet with Russell, whom he had worked with at the White House. Beck did, and Russell encouraged the company to reach out to environmental groups he believed were amenable to an approach.

"What you want to do is to identify the groups that are in the middle, particularly environmental groups that work on a regular basis in Sacramento on legislation where they have strong views that they promote all the time," Russell advised them. "They're used to the give and take of the legislative and political process, so you can reason with those people." Russell's role was to help company representatives and environmentalists "get to the table." Promoting recycling was something the environmental groups and the company could agree on.

Russell connected with Stoddard, whose legislative affairs office in Sacramento was highly regarded and who possessed his own strong relationships with environmental leaders. They met early on with influential environmentalists Mike Paparian of the Sierra Club and David Roe of the Environmental Defense Fund. The groups initially questioned the company's motives for meeting with them. "If your goal here is to try to make nice with us to get some PR that suggests that we're on your side, you're wasting your time," Russell recalled one group member saying. But they were interested in working with the company if it made environmental sense, and recycling made perfect policy sense to them. They wanted to see more communities launch recycling programs and more participation take place. And, of course, they wanted less waste in landfills. The environmental groups' support was critical to upcoming legislation that Waste Management helped introduce.

Stoddard saw the need not to just focus on recycling. He had much bigger plans. He wanted to overhaul California's entire waste management law and undertook a comprehensive rewrite of the state's solid waste statute. The times were calling for it. He had the support of environmentalists, the industry, and progressive pols. Stoddard took his draft legislation to Byron Sher, head of California's House Environmental Quality Committee and a Harvard-trained lawyer and Stanford Law School professor. Sher embraced the idea and ran with it. The bill called for the elimination of the state's existing solid waste law and creation of a six-member Integrated Solid Waste Management Board. Stoddard worked behind the scenes to shore up support for its passage.

The legislation was called AB939. It had teeth. AB939 required every city and county to undertake waste reduction and recycling. Local governmental units had to have state-approved solid waste plans that they implemented and reported on annually. The legislation also set goals to divert waste from landfills: 25 percent by 1995 and 50 percent by 2000. Failure to comply subjected local governments to fines of $10,000 a day, a huge sum. Some local governments opposed the bill, fearing they couldn't comply with its tough provisions. Even so, the bill, slightly amended, was passed and signed into law by Governor Pete Wilson in 1989. Wilson gave the industry's seat on the board to Rosalie Mule, a Waste Management community relations lead in Southern California. Many other states followed the AB939 legislative model.

"That was the turning point in how solid waste would be managed in California and half of it would no longer go to the landfill," Stoddard said. "The company was with us the whole time." Many of the company's internal critics, responding to the clamor for recycling in their communities, came to embrace it. "We're going to make this happen and amend our contracts," Stoddard recalled of their attitudes. "We'll expand our business, it's the right thing to happen, and we're going to make money while we do it."

BRINGING ORDER TO RECYCLING CHAOS

In 1989, Rooney brought Jane Witheridge, then the district vice president overseeing landfills and transfer stations in Pennsylvania and New Jersey, back to Oak Brook to take over the company's corporate recycling effort. By this time, more than 300 local company divisions had begun collecting recyclables. Local managers were responding to local demands and increasingly tougher state legislation. The number of contracts soared: The company had 142 municipal recycling contracts, yet no one had a good handle on what the costs were or where they were. That became Witheridge's focus.

"I did a deep dive," Witheridge recalled. "What do we have, what do we know, what does it cost, and what do we need to do? After interviewing all the people who were associated with recycling that I could find in the company, nobody really knew the costs. There was no standardization, there was no financial system to assess how well we were doing with recycling. People didn't really understand that it was what I would call an entirely new business, which was the separate collection of a product, the manufacturing of it into something else, and selling it."

The financial results of recycling were being reported on different lines in division profit and loss statements. There was little accounting clarity for the collection, processing, and commodity sales.

Hulligan gave Witheridge his marching order: Make order out of the chaos. "It was like drinking out of a fire hose. It was incredibly difficult," Witheridge remembered. "People in local divisions were being asked to respond to local bids for recycling, and nobody knew what to do," she said. "The group that was there was trying to respond in the best way that they could without any knowledge base of what did this mean in terms of return on investment, what should this really look like, and what's the overall profitability. It was just, we have to do it. We have to do it."

Under Witheridge's leadership, the corporate team began to understand recycling's various costs using activity-based costing. Dan Kemna, a recycling director, and Witheridge's "right-hand guy," took the lead in sharing that information with local managers. The group also created what it called MRF (Material Recovery Facility) University to share its knowledge of recycling costs, equipment, and best practices with employees.

NEW EQUIPMENT AND NEW FACILITIES

As the 1990s began, Waste Management also needed to overhaul its equipment that collected recycling. The lightweight compartmentalized trucks widely employed to collect the recyclables were found to be inefficient. Collection was just too slow. During a meeting with Hulligan and Jerry Girsch, one manager stood in a conference room and took a pile of recyclables and sorted them into separate bins. He stated the obvious: "This takes a long time to do if you're standing at the curb and putting plastic and glass and paper in three or four separate bins. It takes forever and you don't get a payload."

A new system was called for. The company phased out curbside sorting and began implementing single-stream recycling, a process that was easier both for the public and for collection and processing. Common today, the system allowed recyclables to be commingled and deposited in a single compactor truck and transported to a material processing facility. Collection then evolved from trucks with a series of bins for recyclables into split body trucks that hauled waste and recyclables simultaneously to larger compactor vehicles dedicated solely to recyclables collection.

Recycling volumes were still growing. In 1990, the company's Recycle America and Recycle Canada programs collected more than a million tons of recyclables, including 781,000 tons of paper fiber, 167,000 tons of glass, 42,000 tons of aluminum and metal, and 20,000 tons of plastics. Two years earlier, it managed 650,000 tons of recyclables.

The volumes continued to skyrocket, and so did contamination, which could affect the value of recyclables on the commodities markets. Cities with contracts that allowed residents to recycle for free but required them to pay for waste collection found residents filling recycling bags with garbage. In Chicago, the company's major blue bag collection program was impeded by complaints of contamination as non-recyclable items turned up in the bins.

Almost every town needed education on recycling. The company produced brochures and fliers and ran newspaper ads on recycling pointing out what commodities could be recycled. Witheridge led a national public relations campaign promoting recycling. Branding it as "Recycling in the '90s," the campaign covered the shared responsibilities of citizens, recycling providers, government, and end-users. It covered recycling's economics and costs and the need to stimulate markets for the material. Recycling wasn't recycling until it found its way into new products, she said in frequent appearances on national news networks. The company also aired a commercial nationwide that featured a young girl educating her parents on recycling while they reacted to neighborhood peer pressure and rushed to put out their bins.

The company's commitment to recycling helped it overcome many of the problems of establishing a new service. Customers welcomed it. They would accept exposure to the commodities markets over the internal protests of some company salespeople who were working to secure new collection contracts and sidestep customer demands.

Waste Management also helped establish facilities to sort and process these high volumes of recyclable materials. In 1991, the company had more than eighty material recovery facilities (MRFs) in operation, which separated corrugated and mixed papers, aluminum and metal cans, and glass and plastics and then baled them into products ready for sale and shipment to end-users to make new products. The manufacturing plants processed more than 1.8 million tons of material on average per year.

MRFs came in a variety of sizes depending on their location and the collection operations that fed them. They could range from processing fifty tons a day to several thousand. The larger facilities incorporated a network of conveyors. The recyclables were separated using a combination of gravity to collect papers, magnets to capture steel, eddy systems to collect aluminum, and optical sorters to separate plastics and glass. The plants were a maze, Rube Goldberg–like systems that worked, but they also relied on people to hand sort and remove items not salable. More than three hundred Waste Management divisions fed the MRFs in 1990, and the search for ever-increasing processing technology was as important as ever. The company's operations in Sweden contributed BRINI automation technology to increase MRF capacity to process comingled recyclables. Kemna, who had a background in the Swiss technology of mixed-waste processing, was instrumental in designing a number of the processing facilities.

In 1995, Waste Management added more MRF capability by acquiring Resource Recycling Technologies, Inc., an owner, operator, designer, and builder of recycling facilities, and New England CR Inc., an operator of fifteen processing plants. By the mid-1990s, Waste Management had more than 170 MRFs in its network, and its hundreds of divisions had accepted and adapted to the business of recycling. The regions and districts had learned to make it work.

At the same time, customers wanted to be confident that their recyclables would find their way to end-users, particularly when commodities markets fluctuated and prices weakened during a steep recession in the early 1990s. The company needed the same assurance. It entered into a series of joint recycling ventures, including the Plastic Recycling Alliance with the DuPont Company to recover plastics found in homes; the Stone Container Corporation to market fiber from old corrugated containers, newsprint, and other papers; and a similar program with the American National Can Company to process and sell aluminum, steel, bimetal cans, and glass containers. Waste Management had become North America's largest recycler. Recycling was now and forever would be a fundamental part of its business.

TEAM MEMBERS LISTED IN THIS CHAPTER

Ron Baker	John Kehoe
Dean Buntrock	Jim Koenig
Jerry Caudle	Al Kromholtz
Bob Damico	Frank McCoy
Jerry Dempsey	Steve Neff
Ken Evans	Kai Nyby
Paul Feira	Jim O'Connor
Don Flynn	Ray Patel
Charles "Mickey" Flood	D.P. "Pat" Payne
Herb Getz	Jay Rooney
Rodney Gilbert	Phil Rooney
John Groenboom	Stan Ruminski
John Goody	John Sanford
Mark Hepp	David Schmitt
Wayne Huizenga	Jerry Seegers
Bill Hulligan	Kevin Stickney
Art Ingram	Jim Teter
John Johnson	Jim Trevathan
Royal Johnson	Alexander Trowbridge
Harold Jorski	Jim Wood

20

THE CHALLENGING EARLY 1990S

By the early 1990s, Waste Management's growth had been accelerating for twenty years. In 1991, Waste Management's revenues reached $7.55 billion—a record high.

Growth was expected, particularly among its investors who had been rewarded because of it. The company had spent much of the past ten years acquiring more technologies and capabilities beyond its waste collection roots to fuel this growth. The acquisitions though—coupled with a steep recession in the early 1990s—would challenge the company's very structure.

LOCAL MARKET OUTREACH

Since its inception, Waste Management had sought growth wherever it might find it.

As early as 1983, it had launched the Waste Management Partners program, forming joint ventures with local garbage haulers in rural America. Opportunity could be found away from the larger markets, and this was seen as a way to extend the company's reach in smaller markets. Don Flynn had persuaded Jerry Seegers, a retired partner from Arthur Andersen, to come aboard. Seegers arrived as an aide to Buntrock, who quickly assigned him to launch the new Partners program.

"Dean Buntrock wanted to participate in smaller communities where he didn't think a community was large enough to support a regular Waste Management operation," Seegers remembered. "So we came up with the Partners program where we bought their customer lists and paid them a fee to run it." Seegers enlisted attorney Herb Getz, then a young associate at the law firm Bell, Boyd & Lloyd, to help, and together, they developed the legal documents and other tools that established the program, Seegers said. Art Ingram joined the program as controller. John Johnson worked to identify prospects, and company veteran Stan Ruminski provided managerial and equipment support to the new companies.

The Partners program offered local operators help in specifying equipment, financing, route planning, safety, sales, and employee relations. Waste Management received customer accounts. In its first year in 1983, twenty "partners" signed on in towns such as Almont, Michigan; Fontana, Kansas; and Waukon, Iowa. By the end of the following year, there were 41 such arrangements in 51 markets in 24 states. The program continued to grow from there. Seegers's team researched local businesses across the US, helping the parent company increase the pace of acquisitions. It was all about growth. "We put together a volume of information about every significant town in the United States detailing who the hauler was and what kind of volume we thought it was doing, which we handed over to Waste Management of North America," Seegers recalled. The Partners program was eventually absorbed into Waste Management's solid waste business.

NON-WASTE ACQUISITIONS

In 1986, its fifteenth year as a public company, the hunt for acquisitions was as energetic as ever. Around this time, the company also started gathering businesses outside its core area of waste. It collected a water services provider, Envirotech Operating Services, in 1986. It started marketing portable toilets, a "Wayne Huizenga idea," Buntrock recalled, in the late 1970s. It added a mobile offices company, Modulaire, in 1987, and temporary power poles and construction fencing services in 1987. These latter capabilities, sold through its local waste divisions, enhanced its sales of big-box industrial containers to the construction industry.

In 1988, Chem Waste entered the asbestos abatement and scaffolding business through its acquisition of 49 percent of the publicly traded The Brand Companies. By the late 1980s, the company had formed a unit called Urban Services, which was created out of the TruGreen lawn care company (acquired in 1988 from Huizenga and partners) and a series of pest control business acquisitions. Seegers led the Urban Services unit, which was traded to publicly traded Service Master in 1990 for a stake in that company.

But one of its most significant transactions occurred in 1988 when Waste Management contributed its Tampa McKay Bay waste-to-energy plant, Envirotech water business, and engineering resources to a new enterprise called Wheelabrator Technologies Inc, created by the holding company, The Henley Group.

THE TRASH-TO-ENERGY CROWN JEWEL

Wheelabrator was a technological "tuck-in" that gave Waste Management's solid waste business an additional disposal dimension: burning trash to create energy. As the nation's leading trash-to-energy enterprise, Wheelabrator's primary business was developing and operating waste-to-energy plants. With Waste Management's Tampa operation, it included 14 plants pumping out 12 million kilowatt-hours of electricity a day—the equivalent of 17,000 barrels of oil. It had plants in Connecticut, Florida, Maine, Maryland, Massachusetts, New Hampshire, New Jersey, Pennsylvania, and Washington state with a combined output of 575 megawatts. Wheelabrator was the sixth-largest non-utility power producer and was planning for even more plants.

Wheelabrator had a long history. The public company was itself a product of mergers. The original company, called the American Foundry Equipment Company, dated back to 1911. It changed its name to The Wheelabrator Corporation in 1932 and became Wheelabrator-Frye Inc.

in 1968. The company was acquired by The Signal Companies in 1983 and became publicly traded in 1987.

Phil Rooney and Jim Koenig led the negotiations with experienced dealmakers Paul Meister and Paul Montrone representing Henley. The transaction granted Wheelabrator access rights to certain Waste Management disposal sites to develop new facilities there and dispose of its ash. In return, Waste Management received a 22 percent stake in Wheelabrator. The relationship with Wheelabrator proved successful, and by 1990, the company increased its stake to 55 percent—a majority, and control. It was a complicated transaction.

"We now had to deal with another majority-owned publicly traded company, Wheelabrator, that was in the waste business, which could very quickly lead to conflicts of interest about business opportunities and claims by minority stockholders of Wheelabrator of unfair dealing," recalled Herb Getz, the assistant general counsel who played a key role in all events transactional. "This required all companies to get something in exchange and approval by multiple boards of directors, and by Wheelabrator stockholders." To resolve the matter, intercompany agreements were created allocating business opportunities among Waste Management, Wheelabrator, and Chem Waste, which was also involved.

Koenig and Getz spent several days in New York negotiating a letter of intent for the 1990 transaction with Meister. They returned to Oak Brook on a Friday evening expecting Buntrock and Rooney to approve it. Any congratulations for their efforts were fleeting. Rooney ordered the two back to New York the following Monday to negotiate a definitive agreement. He wanted the deal announced quickly. The two reminded Rooney that they were supposed to go on a long-planned outing to the famed Augusta National Golf Club with Lee Morgan, a company board member and former chairman and CEO of heavy equipment maker Caterpillar, Inc. "Rooney said he understood that but had already given our invitations to Jerry Girsch and Bill Hulligan," Getz recalled. The two went to New York, hashed out the agreement, and the merger was announced.

Buntrock had also embraced waste-to-energy. It had been part of his vision from the beginning, particularly since landfill development was a difficult proposition. Community opposition was always stiff. Many millions of dollars could be invested in a landfill project in land expenses, engineering, and consulting and lawyering with no guarantee of success. Many "greenfield" projects, those proposing new sites, could go on for years.

With Wheelabrator now as a subsidiary, Rooney wore two hats: president and chief operating officer of Waste Management and chairman and CEO of Wheelabrator. Rodney Gilbert was Wheelabrator's president during its first years within Waste Management. He soon moved on to become president of Rust International, which was formed in January 1993 during a reorganization of Chem Waste and Wheelabrator's environmental engineering, construction, environmental consulting, and remediation services businesses.

The day-to-day operating responsibility for Wheelabrator then fell to John Kehoe, a smiling, mustachioed executive who bore a passing resemblance to Teddy Roosevelt. With his larger-than-life bearing, he made a strong impression and was adept at communicating his upbeat— he would say "sanguine"—story to investors. He had been a paratrooper in the Army's 82nd Airborne Division and spent ten years in sales and marketing at IBM.

Kevin Stickney, who headed Wheelabrator's communications group, recalled the time he persuaded a *Wall Street Journal* reporter to come up to Wheelabrator's headquarters at Liberty Lane in Hampton, New Hampshire, to interview Kehoe. Stickney's coup turned comical. Kehoe thought himself the spitting image of actor Tom Selleck of the popular Hawaii-based TV show *Magnum P.I.* To emphasize the resemblance, Kehoe used to wear a Detroit Tigers baseball cap just as Selleck had on the show. On being introduced, Kehoe asked the *Journal* reporter if he reminded him of anybody. The journalist thought for a moment but couldn't quite put his finger on it. Keep thinking, Kehoe told him. Finally, the reporter came up with it: "I know, Wilfred Brimley." (The much older Brimley had played roles in such movies as *Cocoon* and *The Natural*, a far cry from the younger Selleck.)

Kehoe was, however, a serious manager with strong people around him in operations, finance, engineering, and development. Among them were Jim Wood, who directed plant operations; Ray Patel, who guided the company's engineered systems unit; and John Sanford, his CFO. Mark Hepp ran engineering activities, and David Schmitt was general counsel.

Wheelabrator offered more than trash-to-energy technology. It had air pollution control systems and water and wastewater treatment capabilities. Overnight, the Wheelabrator investment made Waste Management an international one-stop-shop for a range of environmental services.

Wheelabrator's Clean Air Group, led by Paul Feira, served the utility, steel, foundry, pulp and paper, cement, and waste services industries. John Goody directed its Clean Water Group, which provided technology, products, and management services to water and wastewater treatment operators. And the former Wheelabrator Rust Engineering Group, which became for a time a publicly traded Rust International subsidiary guided by Gilbert, had customers in the pulp and paper, aerospace, manufacturing, food and beverage, metals, and chemicals industries, as well as government.

With Wheelabrator in its fold in 1991, Waste Management celebrated its twentieth year as a publicly traded company. As its scale and reach expanded, company executives continued to search for ways to sustain its historic rate of growth, a key metric needed to keep Wall Street's goodwill and continued investment. Waste Management's size was now such that sustainment would be harder, much harder to achieve. Every sizeable acquisition in the solid waste business was under the Justice Department's Antitrust Division's microscope, and the company had already established strong positions in most larger markets. There would still be acquisitions aplenty, but the executive team also needed to generate more revenue internally. The pressure was on.

A NEW SALES STRATEGY

Until the 1990s arrived, sales in the solid waste business, which then accounted for about 60 percent of its total revenue, had been the sole purview of the local division manager. In some instances, the function was seen as unnecessary as new accounts arrived magically and many calls came in unsolicited from customers. Typically, however, a division sales manager directed a team of four or five people, depending on the location.

Most salespeople were drawn from operations and other division functions. They each had a territory. They were hardworking, knew the garbage business, and knew their accounts, their collection schedules, and the details of their customers' requirements. What size containers did they use and need? Compactors? Rolloffs? Collection frequency? The sales reps would work the phones, drive their standard Dodge Dynasties to prospects, and wear away their shoe leather knocking on doors to drum up new accounts. They maintained relationships and also communicated price increases. The best sales prospect, they knew, was the one found between the two customers whose containers Waste Management was already collecting. The sales reps might be assisted by an inside person who put together sales contracts and provided other support.

Still, better sales results, more focus, and more customers were needed for the company to sustain its growth. Rooney wanted better performance. The company needed it. Rooney turned to an executive outside the industry from Big Blue, IBM, the huge computer maker, which then generated nearly $70 billion in sales. D.P. "Pat" Payne, IBM's former Midwest general manager, joined Waste Management in 1990 as a senior vice president and the company's first chief marketing officer. Payne had grown up in Waco, Texas. After graduating from Texas A&M in 1964, he went right into the Army and was a Vietnam veteran. At Waste Management, Rooney assigned Payne responsibility not only for sales and marketing but state government affairs, information systems, human resources, and quality programs as well. His arrival freed other executives to focus on operations undistracted by administration.

This was not the company's first such effort to improve sales. Stan Ruminski had been handed that job in the mid-1980s. He, in turn, hired a sales consultant, Al Kromholtz, to put together a sales system that could be replicated across all local divisions. Its objective was to professionalize selling and develop the common message that although Waste Management was a national company, its operations were local. At the time, the trucks still often carried their former owners' division names. Local identification had value.

Ruminski and his team, including Kromholtz, Kai Nyby, and Steve Neff, went on the road pitching sales best practices. Kromholtz promoted the "ABCs" of selling: "Always be closing." Watch for the customer's cues, aim to close, Neff recalled. "Kromholtz would emphasize, 'If you identify a closing signal, stop talking, listen, and close.' He made a great deal of sense since most professional salespeople talk too much." Too much talk was a turnoff. Closing counted.

In joining Waste Management, Payne quickly recruited other IBM colleagues to the newly conceived corporate and field sales function. Ken Evans came in as vice president as did other IBMers in area and corporate roles.

The new class of salespeople was doing its best to raise the quality of the existing sales team, build on a nascent national accounts program launched only months earlier, and generate better results. It would be a challenge. They were not selling mainframes. Selling garbage services, a somewhat simpler concept on paper, was often more complicated in reality. It was one prone to stiff competition, aggressive pricing, and the vagaries of geography. Connections were often made at the loading dock, not the corporate conference room. Some of the most proven and productive salespeople departed under the new system, sometimes under pressure, other times failing to conform to the new team's shifting approach and culture.

National accounts contracts were additive when secured. But often, the company's local hauling divisions already included the local operations of many big national companies as their customers. Competitors had a share of the local operations of national corporations and wouldn't give them up easily. And, of course, the customers had their baked-in loyalties. Getting them to switch haulers was not easy. Some of the most experienced waste operations executives in the regions failed to find wisdom in the new top-down sales approach. Others resented it. A number resisted. "How could a bunch of yellow-tie, blue-suited, short-haircut, Ivy League boys come in and tell us how to run our business?" one sales veteran remembered. A few longtime managers were more blunt: "They didn't have a clue."

The first big national account the new sales team secured was Walmart, a company notorious for beating down suppliers to low prices. The sales team faced off against the disciplined Walmart people in a barren Bentonville, Arkansas, room with a linoleum tile floor where many had hung and displayed their wares in hopes of the retailer selling them in its stores. A retired Waste Management employee recalled how, at one meeting, the Waste Management team sat in chairs behind a table pressed closer to a wall, allowing the Walmart people more ample room to negotiate. They whittled away at Waste Management's pricing until, finally, the senior salesperson said they had no choice but to give up the contract because it was a money loser. Only then did the Walmart team show mercy and allow an acceptable price. Walmart would be the first of many national accounts won.

The new corporate marketing team succeeded in installing more sophistication into the business. It was a culture change. Every region now had its own sales manager overseeing a more local manager. Training was upgraded and mandatory. Technology common today was introduced. Call centers and phone systems with tree structures now guided customers to specific services or to complain, perhaps, about a missed pickup. Some salespeople specialized in certain accounts. The new salespeople were trying to elevate the function.

"They actually brought in a number of great supportive strategies to our programs," said Neff, a former teacher hired into a sales position in 1979 by veteran Chicago-area garbage managers Frank McCoy and

John Groenboom. "Information management systems were improved. Metrics were introduced for accountability and sales tracking. Sales reporting from each salesperson flowed up to sales managers and onto sales directors," he said. "They brought accountability that could be measured and managed instead of the former wild west approach to let sales do whatever they do and let operations take care of it."

A SLUGGISH ECONOMY

However, it wasn't the best time to change a business model and find growth. The 1980s had been a boom period. Business and life were good. In fact, the summer of 1990 was reportedly the tail end of the nation's longest period of peacetime expansion. That changed with the recession of the early 1990s. Pundits called the period "sluggish." In 1992, the weakened economy helped sweep George H. W. Bush out of the White House and Bill Clinton in.

The fluish economy had several causes: The Iraqi invasion of Kuwait in August 1990, though short-lived, sent oil prices soaring. The Federal Reserve adopted a tight monetary policy trying to prevent the economy from overheating, which limited the nation's growth. Real estate values were slammed, the stock market was turbulent, and unemployment rose.

Waste Management suffered the economy's stinging impact. The next several years were challenging, to say the least. Things slowed. The company's results were a "lagging indicator"—the slowdown seeped predictably downward to the company after major industries first experienced less consumption and production. The economic downturn meant less waste, solid and hazardous, to manage.

The 1991 report to shareholders, usually a confident communique, could not avoid referencing the distress while also conveying hopeful optimism: "The past year was extremely challenging, the recession deeper and more persistent than was projected. There was weakness in the base business and remediation services of Chemical Waste Management, and the recession affected our industrial customers who use our solid waste container services, including those in the construction industry.

Still, Waste Management continued to grow, and to grow impressively." Company leadership expected the weakness "to linger." It would.

In 1991, North American solid waste revenues grew 8 percent, far from the normal double-digit growth rate. The company served 12 million households and one million businesses. Chem Waste revenues rose 18 percent to $1.4 billion, but its profits declined. Wheelabrator profits grew in its trash-to-energy, clean water and clean air, and engineering units. And Waste Management International revenues rose 31 percent as its global acquisition spree continued.

The annual message to shareholders describing 1992 was even more direct, reporting: "Over the years, we have achieved much success, and 1992 saw that positive momentum continue. But frankly, even while we built on our leadership and expanded our business, the last year was a pretty difficult time. You have heard a lot of companies saying this. In our case, the poor economy put terrific pressure on our industrial and commercial customers and slowed the growth all of us have come to expect."

The domestic solid waste business saw weak prices for recyclables, which had grown to be a major component of its operations. Chem Waste's customers cut production and deferred spending on cleanup jobs. Only Wheelabrator's revenue rose as it added capability in its clean water and clean air divisions. The new Rust International engineering subsidiary was created by carving out pieces of Chem Waste, Wheelabrator, and The Brand Companies asbestos abatement and scaffolding business. Rust also had resources dedicated to engineering, environment and infrastructure, remedial services, and industrial and construction services, as well as an international component.

In 1992, Waste Management's board, recognizing its vastly expanded capabilities and what many of its executives believed its mission now to be, changed the company's name to WMX Technologies—a global environmental services company. It would no longer be known only as a garbage company. It had become an environmental mutual fund, one industry analyst thought. "Our customers' needs have grown for more sophisticated environmental services and solutions," management told shareholders. "We have put into place a collection of technological and scientific resources unequaled in our industry. These resources for the environment extend far beyond our original waste services business."

WMX Technologies had, indeed, a lot of capability and a lot of add-ons. With the ailing economy fueling its poorer results, Buntrock and Rooney launched a cost-cutting program in 1992 that trimmed staff and streamlined management. The North American garbage business, now representing half of revenues, was realigned from nine to four regions, with Jerry Caudle heading the West, Bill Hulligan the Midwest, Charles "Mickey" Flood the East, and Jim O'Connor the South. They expanded the customer service and sales functions and introduced more automation in hauling units. More efficiency was needed, and costs had to be cut.

Pat Payne was put in charge at Chem Waste, succeeding Jerry Dempsey, an ex-Borg Warner executive and WMX Technologies board member. Dempsey became Chem Waste's chairman. Payne shifted Chem Waste's organization to align geographically with the solid waste operations, committed more resources to improve customer service, and streamlined waste acceptance processes. Responsibility for profits was moved from the operating units to headquarters. The re-sizing fit the business's reduced scale.

1993: A DIFFICULT YEAR

Despite these moves, 1993 wasn't any easier. There was less garbage to collect and more disposal space to maintain, although projects to clean up the debris from Hurricane Andrew in the year's first half helped. Recycling markets were called "stagnant." WMX's leaders again invested in sales. Their desire to put the most experienced decision-makers closer to the customer triggered yet another organizational realignment. The four regions of only a year earlier reverted to nine headed by well-seasoned managers: Jim O'Connor, Florida; Mickey Flood, Mid-Atlantic; Ron Baker, Mideast; Bill Hulligan, Midwest; Bob Damico, Mountain; Jay Rooney, Northeast; Harold Jorski, Southeast; Jerry Caudle, Southwest; and Jim Teter, West. Five hundred positions were eliminated.

The Chem Waste corporate staff, meanwhile, had grown larger. It had layered on people in sales and research and development roles. The company had expanded during a time when the poor economy had masked a significant reversal in the hazardous waste marketplace, one not foreseen or diagnosed. Chem Waste's woes weren't just due to the economy. The regulatory environment created uncertainty for generators. Regulations reduced hazardous waste volumes and postponed Superfund cleanup projects. Stiff taxes on out-of-state wastes hit Chem Waste hard in Alabama and Louisiana. With industry overcapacity, the company's incinerator and disposal sites suffered.

An earlier August 1992 *Wall Street Journal* story reported on some of the problems Chem Waste faced. Headlined "Economy Alone May Not Rejuvenate Chemical Waste," the story noted that contaminated site cleanup work was increasingly completed at the site, precluding the shipping of hazardous dirt and debris to company disposal locations. The article reported, "Now with signs of economic recovery emerging, the company's followers are wondering if the business will return to its lucrative, pre-recession levels. Only partly, it appears, because some fundamental changes in the hazardous waste industry are also unfolding that could reduce profitability for years to come."

Chem Waste needed answers.

A SURPRISING HAZARDOUS WASTE DISCOVERY

In 1993, Buntrock dispatched Alexander "Sandy" Trowbridge, a company board member, the former secretary of commerce in the Johnson Administration, and a retired president of the National Association of Manufacturers, to organize a tour of major corporate customers and report back. Buntrock led Dempsey and Trowbridge on the first ten visits. They were accompanied by Neff and later, Jim Trevathan. Trowbridge packed the schedule. The travelers jetted out before sunrise and reached several companies in a day. They visited about fifty companies in just six weeks. The Trowbridge team sat down with only the most senior executives—CEOs and the "numbers two and three" in corporate hierarchy—people who could demand and get answers from their manufacturing chiefs on their plants' hazardous waste generation.

Trowbridge's assignment was part information-gathering, part marketing. The tour team's visits to the corporate suites worked to impress the executives with WMX Technologies' expanded environmental portfolio and its mission to be one of the most important environmental companies in the world. Trowbridge introduced the team, and Buntrock and Dempsey reviewed WMX's two decades of growth. Neff outlined WMX's broader capabilities, paging through a flip-the-page presentation booklet containing 18-by-24-inch illustrations and charts.

What the Trowbridge team learned was stunning: They discovered that their customers' needs had changed. In their efforts to reduce costs during the recession, the companies had grown wiser about regulations and smarter about their own manufacturing processes. They had reengineered their systems and, in the process, reduced their hazardous waste generation and their costs. Those wastes weren't coming back, at least not in the volumes they once had.

"Lo and behold, they didn't need to use hazardous materials," said a retired senior executive. "I'd say that 70 to 80 percent of the big waste streams that these companies were producing went away overnight. In a very short period of time, these huge figures that the government was working off of in their [hazardous waste] surveys just went away."

Companies had recognized the costs of managing their wastes and directed their engineers to find ways to reduce their hazardous waste generation, recalled Royal Johnson, a longtime financial executive at Waste Management and Chem Waste. "They would engineer their way down so the volume would be reduced. A lot of the demand that everyone thought we were going to have was going to go away." Industry incinerator overcapacity slammed the company, too, Johnson said. "The demand for hazardous wastes services went down so then obviously the pricing started dropping dramatically because you had too many people chasing the same wastes."

Chem Waste had no choice but to act. In 1993, it restructured its business, wrote down assets, and took a $363-million hit to earnings. Sixteen hundred people lost their jobs. It was bitter medicine to swallow. In early 1995, WMX Technologies bought back the public shares of Chem Waste, making it a wholly owned subsidiary again. It also bought Rust International's public shares later in the year. The company was working hard to simplify its structure and solve its complexity. In 1997, it purchased Wheelabrator's public shares, too. Only Waste Management International would remain a publicly traded subsidiary.

For WMX Technologies, the early nineties were a difficult period. The pressure was on. And the next half-decade would not be any easier.

TEAM MEMBERS LISTED IN THIS CHAPTER

Larry Beck	Larry O'Donnell
Dean Buntrock	Peer Pedersen
Don Chappel	John Pope
John Drury	Rod Proto
Jim Fish	Colleen Quenzer
Larry Galek	Robert Reum
Herb Getz	Phil Rooney
Rod Hills	Steven G. Rothmeier
Wayne Huizenga	John Sanford
Joe Holsten	John Slocum
Bill Hulligan	Tom Smith
Robert S. "Steve" Miller	David Steiner
A. Maurice "Maury" Myers	Bill Trubeck
Ronald LeMay	Ralph Whitworth

21

THE RETURN TO WASTE
MANAGEMENT'S ROOTS

The year 1996 marked a milestone for Waste Management: It celebrated the twenty-fifth anniversary of its debut as a public company. The business that had started with just a little more than $10 million in annual sales was now a global multi-line environmental services company known as WMX Technologies, Inc., with annual revenues totaling more than $10 billion. It had grown from four hundred employees to more than 73,000 team members.

Founder and CEO Buntrock had worked for twenty-five years to assemble a collection of environmental and infrastructure businesses operating in more than twenty countries. The breadth of services its companies offered was in many ways hard to comprehend. They were organized into five distinct operating groups:

Waste Management, Inc., the traditional North American solid waste operations that now included Recycle America collection and materials recovery operations, medical waste services, special event and construction site services (marketed under the Port-O-Let, Modulaire, and WMI Services names), landfill liner and containment systems (offered through its 51 percent stake in the National Seal Corporation landfill liner subsidiary).

Chemical Waste Management, Inc., a 78-percent-owned, publicly traded subsidiary providing hazardous waste services, and through its Chem-Nuclear Systems business, offering low-level radioactive waste services.

Wheelabrator Technologies Inc, a 56-percent-owned, publicly traded subsidiary primarily dealing in the development and operation of waste-to-energy facilities but whose operations included a dizzying array of other environmental and infrastructure businesses, including clean water operations (EOS, BioGro, and IPS), clean air companies (Wheelabrator Air Pollution Control, Altech, Pullman Power Products), VOC companies (ARI, Huntington Energy Systems, Westates Carbon) and the Wheelabrator Engineered Systems group (screen products, water process systems, and surface preparation equipment).

Rust International Inc., jointly owned by Wheelabrator and Chem Waste. In addition to its long-standing Birmingham, Alabama–based industrial engineering and construction operation, Rust's portfolio also included environmental engineering services (Rust Environment & Infrastructure), hazardous waste remediation services (which included the former Chem Waste ENRAC operations but were then offered through its publicly traded OHM Corporation subsidiary), asbestos abatement services (offered through another publicly traded subsidiary, NSC Corporation), and the industrial scaffolding, cleaning, and plant services operations previously offered through The Brand Companies Inc. subsidiary.

And, finally, Waste Management International plc, its London-based international operations that offered the full range of solid and hazardous waste management services in seventeen countries.

The company's scale and breadth of services were so broad that it could be easy to lose sight of how far its leaders had taken it from its roots on Chicago's South and West sides. The pace of change, and the challenges of managing such a business, gave the management team little time to reflect on what had been accomplished. With an enterprise of this size and scope, opportunities and challenges presented themselves every day. And now they would face new and unexpected challenges.

SUSTAINING UPWARD GROWTH

Buntrock had led WMX Technologies' growth, historically 15 to 20 percent a year. Investors loved this growth as their holdings grew. Why wouldn't they? The company had assembled impressive assets, but its management recognized that continuing to grow at this rate was no longer realistic. Like the laws of gravity, finance's "Law of Large Numbers" meant achieving past results and growth rates would be more and more difficult. And any company, particularly one of WMX Technologies' scale, would not be able to sustain its historic growth rate. After having achieved $10 billion in 1996, the company would need another $1.5 billion a year in new business to sustain its 15 percent rate. That was an unlikely prospect.

The senior management team began a formal review of its businesses and their prospects to come up with a plan. They employed business and financial consultants to help, including Andersen Consulting and Merrill Lynch. Early conclusions were drawn. The company, searching for solutions, adopted what it called its Expanded Management System, a cultural change initiative made up of four elements, or "The Four Pillars," as those inside the company called them: customers, employees, the environment, and shareholders. The idea was that satisfied customers served by committed employees doing the daily work of collecting the garbage and protecting the environment would produce shareholder value. The system became a mantra. It might help to control some costs, but it wouldn't generate the means to grow substantially.

The mid-1990s brought more clarity, but the challenges persisted. Management committed to aligning the company's capabilities in hopes of capturing the most likely growth opportunities. They simplified the organization into four global business lines: waste services, clean energy, clean water, and environmental engineering and consulting. They focused on internal growth, not acquisitions, increased returns on assets, and more cash generation. The company also reduced its publicly owned subsidiaries to two, Wheelabrator Technologies and Waste Management International. The board authorized the repurchase of 25 million WMX Technologies shares.

NEW LEADERSHIP

With twenty-five years of leadership behind him, Buntrock determined that it was time to hand the CEO reins to his longtime partner, Phil Rooney.

Rooney had come into the company as a young man fresh out of the Marine Corps, a decorated soldier in Vietnam. He knew and loved the business, and indeed, had grown up with it. His career had begun in the company's original garbage truck yard in Cicero, Illinois, where he first met and impressed Buntrock. After rising up the ranks in various senior officer roles, the board elected him president and chief operating officer in late 1984. Nearly a dozen years later, in May 1996, at age fifty-one, he was promoted to the company's top job as chief executive officer. Buntrock had planned to retire as CEO that June and remain chairman for another two to three years.

Rooney was on a first-name basis with managers in the divisions and employees in the trenches and on the trucks. Those in operations thought him their trusted connection to the corporate office. He was as sophisticated as he was smart. He could be a no-nonsense manager demanding performance when getting results could be difficult. He knew, too, how to express his appreciation for a job well done. He was deep into the details of the business and couldn't be fooled when his managers' answers didn't add up. He generated a lot of loyalty. Rooney planned to streamline the company and turn it back to its solid waste roots—a massive overhaul he was determined to undertake. His plans, approved by the board at a February 1997 meeting at the Breakers Hotel in Palm Beach, Florida, included divesting non-core and non-integrated businesses, eliminating jobs, and buying back company stock. It also included changing the company name back to Waste Management, Inc.

But about the same time he was promoted, in the spring of 1996, for reasons that can only be speculated upon, a group of large institutional investors, led by the huge Soros Group of funds and the activist investor Lens Fund, focused on the company—and on Rooney.

OUTSIDE INVESTORS MOVE IN ON WMX

Led by Robert A.G. Monks and Nell Minow, the Lens Fund targeted the management of companies they believed underperformed for shareholders and then proposed steps they thought would unlock shareholder value. They had reportedly looked at WMX Technologies several years before, when they lunched with investor George Soros and suggested moving in on the company. Soros had then passed on it, but now, in early 1996, they decided to team up. Lens's team and Soros wanted a strategy that would ultimately transform the company, returning it to its North American waste management roots. Ironically, it appeared to align with Rooney's plan.

Lens's Minow leveraged her contacts in the media and got attention—and investors—in their campaign. In the spring of 1996, the modest firm invested $25 million in WMX Technologies, according to writer Hilary Rosenberg in her 1999 book about Monks titled *A Traitor to His Class*. Compared with other investors, the Lens stake was a paltry sum given that there were about 450 million shares outstanding, but Lens had ownership and, through it, power.

They continued their campaign through the balance of 1996 and accelerated it in early 1997. Their attacks targeted not only the company's strategy but senior management and the board. It often became personal. The first casualty of this campaign was Rooney himself, who stepped aside as CEO in February 1997.

It was only eight months after he had become CEO, little time to execute an overhaul of the company. Across the enterprise, managers and their staff and crews were in disbelief, "surprised and disappointed," the company's spokesperson said at the time. Rooney's efforts to streamline the company and turn it back to its solid waste roots would be incomplete. "I am not prepared to let personal attacks distract this company," he wrote to the board and the company's employees in 1997. Buntrock returned as CEO for the interim, and the board engaged the Heidrick & Struggles executive search firm to hunt for a new CEO. Within months, the departed Rooney became vice-chairman of ServiceMaster, a company that operated home services businesses.

For Buntrock, Rooney had been a partner from the earliest days. "It is an understatement to say that his leadership and contributions to our success have been enormous," Buntrock told shareholders. "He truly has represented the drive and spirit of our company and, during his twelve-year tenure as president and chief operating officer and then as CEO, the market value of WMX climbed from $2 billion to $16 billion."

Rooney's departure ushered in a series of events and controversies that could not have been anticipated by anyone associated with the company at that time. Emboldened by the results of their efforts thus far, the Lens and Soros investors pushed Buntrock and the company's board to add additional independent directors and to hire a new CEO from outside the industry.

In March, Buntrock recruited two new board members. He was looking for additional financial perspective. The company named Steven G. Rothmeier, a former chief executive of Northwest Airlines, as an outside director, filling Rooney's vacancy, and nominated Robert S. "Steve" Miller to stand for election to the board as another independent board member. Miller was a former chief financial officer of Chrysler Corporation during its crisis days and the 1980 government bailout. He had also been the acting CEO of Federal Mogul Corp. Miller didn't know it yet, but he would soon have his hands full. Peer Pedersen, a Chicago attorney and the senior outside member of the board, led the board search for a new CEO.

A NEW ERA OF LEADERSHIP

On July 14, 1997, the company announced the hiring of Ronald LeMay as chairman and chief executive officer. LeMay had arrived from Sprint Corp., the telecommunications company based in Kansas City, where he had been president and chief operating officer. Buntrock stepped down as chairman but kept a seat on the board. He planned to leave at year-end.

The Soros Fund congratulated the Waste Management board, saying LeMay would be an outstanding chief executive officer. Stanley

Druckenmiller of the Soros firm issued a statement: "Mr. LeMay's impressive track record at Sprint, including his turning around the long-distance business in the early 1990s, make us confident that he has the leadership qualities to realize the value we believe is inherent in Waste Management's business."

"He challenged us a lot," Getz said of LeMay. "He had come from an industry with falling prices, which had to find new ways of doing business more efficiently. At one point, he asked whether there was any reason we couldn't think about acquiring BFI. Everything was on the table. He was looking for an overarching strategy that the A. T. Kearney (consulting firm LeMay engaged) might discern from their work to buy some time to turn the company around."

LeMay seemed uncomfortable with the fit from the beginning, some executives later commented. He had been meeting with top executives at the headquarters office on Butterfield Road trying to get his arms around the company. Board member Miller had regarded LeMay as a class act, a leader with a great resume. He was thrilled to have helped attract him to the company. During the recruiting process and later compensation negotiations, Miller sensed LeMay's nervousness about leaving his familiar post at Sprint and going to a new and unfamiliar industry. LeMay was a telephone man. Managing a waste services company was miles from his experience. Miller thought LeMay should avail himself of Buntrock's industry knowledge and experience.

Then, the day before Halloween, Getz got a call from Colleen Quenzer, his former assistant who was then working for LeMay. "Ron wants to see you," she said. "I'm due at an [OHM subsidiary] board meeting this morning," he replied. "No, you need to come here," she insisted, Getz recalls.

LeMay was operating out of Phil Rooney's former third-floor office in the southeast corner of the main headquarters building overlooking a pond. Quenzer put her head down and shook it slightly when she saw Getz arrive. "My first thought was that it's my last day at work," said Getz. "I go in and close the door behind me. To my relief, Ron started by saying, 'This is going to be the most challenging day of your career. John Sanford [the CFO] is leaving.'" Getz feigned surprise about Sanford, protecting a source who had tipped him off about it the night before. Getz knew that LeMay was taking a closer look at certain accounting issues that he and Sanford had brought to LeMay's attention. But he was shocked by what he heard next. LeMay looked at him and said, "And this is my last day, too," Getz remembers. "He told me, 'If I were you, I would call a meeting of the board.' We shook hands, and I walked down the hall."

Chief Operating Officer Joe Holsten, promoted from his Waste Management International position in February, remembered the day clearly. He had found LeMay "a little bit of an academic" in addition to being likable and level-headed. He knew LeMay was concerned about the company's accounting. LeMay came to Holsten's office down the hall to deliver the news: "I didn't sign up for this," Holsten recalls LeMay telling him. "I'm leaving." The executive team was busy implementing the A.T. Kearney plan LeMay had commissioned. Holsten said LeMay told him that the plan was a good one, and there was nothing he could add that the team wasn't already doing. With that, LeMay boarded a plan dispatched by Sprint and returned to Overland Park, Kansas, where he returned to his former position as president and chief operating officer.

LeMay's concerns about some of the company's accounting practices contributed to his resignation, coupled with a lucrative offer to return to Sprint, several Waste Management executives believed. Miller expressed disappointment over LeMay's departure. He had been preparing to teach a class in crisis management at Carnegie Mellon University in Pittsburgh when he heard the news from LeMay. Getz jokingly told him, "The board has elected you chairman and CEO. That should teach you never to miss a meeting. We've sent the company plane to Pittsburgh to pick you up and bring you to Chicago for dinner with the senior management. You're in charge now." With his experience in similar situations, Miller was pressed into service as acting CEO. (Interestingly, Miller had been in discussions with Bud Selig, owner of the Milwaukee Brewers, to become commissioner of Major League Baseball [MLB]. Selig later informed him that he'd decided to take the commissioner's job on an interim basis. Selig ended up with the job until 2015.)

THE LUMBERJACK CEO

Handing the CEO job to Miller was the logical choice for the board. He came armed with the right stuff and appropriate experience.

Miller, a tall, genial man, was the grandson of a successful timber miller with roots in Bandon, Oregon. His father was a successful Portland lawyer who was instrumental in the growth of the Georgia-Pacific Corporation. He learned valuable lessons as a child visiting the Bandon mill and later working for the family as a young lumberjack harvesting trees that stayed with him. He understood "the dignity of work and the rewards of honest enterprise," he later wrote in his book *The Turnaround Kid.*

Miller had begun his business life in the finance department of the Ford Motor Company in Detroit. His career there led to tours of duty in Mexico, Australia, and Venezuela. Along the way, his boss in Caracas, Gerald Greenwald, had been recruited to Chrysler by Lee Iacocca, the former president of Ford who had been fired by Henry Ford II. When Iacocca rose to become Chrysler's chairman and CEO in 1979, Greenwald moved from controller to executive vice president. During that time, the troubled automaker was on the verge of bankruptcy. Chrysler officials were traveling to Washington to beg the federal government to guarantee a refinancing of the company loans with hundreds of banks and other institutions.

Greenwald was looking for an executive to lead a massive financial reorganization. He found him in Miller, who joined as an assistant controller and rose to be the company's CFO in 1981. The Chrysler challenge he faced could not have been more complicated or had more moving parts. More than $1 billion in loans from hundreds of lenders had to be reworked. The automaker's unions had to agree to concessions. After all of this, Congress had to approve the aid package for the company, which it did in December 1979. President Carter signed the legislation on January 7, 1980. Miller then had to first persuade the multitudes of reluctant lenders to believe Chrysler had a future building cars and, second, to agree to revised and likely less favorable terms. His marathon of meetings worked: Over time, the lenders lined up in agreement. On the day the restructuring was sealed and the money finally in hand, Miller, the architect and chief negotiator of the loan restructuring, walked up to a teller at the Manufacturers Hanover Bank in New York and personally deposited the check for $486,750,000 in a Chrysler account to keep the automaker going. Chrysler ended up borrowing only $1.2 billion and paid it back in seven years. It was the beginning of Miller's career as a turnaround specialist. More challenges would follow, Waste Management among them.

Miller arrived at Waste Management's corporate offices in March 1997. He was an executive unafraid of crises and with a reputation for dealing with tough situations. Highly educated, with Harvard Law and Stanford MBA degrees, Miller was, at the same time, down-to-earth and plainspoken. He puts on no airs. Those were fitting traits to possess as he faced a situation that had come suddenly.

Now in charge, Miller immediately added two more outside directors to the board: Rod Hills, a one-time chairman of the Securities and Exchange Commission, and John Pope, a former president of UAL Corp, the parent company of United Airlines. He quickly communicated with the investor community and with employees. He was working to calm the waters.

With the board now firmly in the hands of Miller and a majority of professional directors with decidedly conservative financial credentials from outside the industry added, the board turned to focus its work on the company's accounting issues, and Miller took on the task of finding a new leader for the company. A new team of Arthur Andersen auditors with no prior experience with waste industry accounting, under the direct supervision of Hills, now chairman of the board's audit committee, conducted a review of the company's financial statements. The results of their review were and remain controversial.

The new team of accountants, with the full support of Hills, recommended a broad restatement of the company's financial statements. These accounting adjustments focused primarily on a number of judgmental areas, including equipment depreciation and salvage policies, as well as landfill cost accounting procedures. The restatement produced no changes in the company's reported revenues or cash flow. Subsequent litigation relating to the restatement saw the company's executives, and their accounting experts, defend the earlier financial statements. They argued that they believed the company's accounting judgments at the time were well supported by the facts and reasonable under the circumstances. The executives criticized the decisions by the new and more conservative team in charge as simply substituting their own judgments, with the benefit of hindsight, for those previously made and reviewed at the highest levels of Arthur Andersen & Co. The matters eventually led to an SEC complaint against Arthur Andersen and several executives. Ultimately, the matters would be resolved with penalties being paid by both parties and neither the executives nor Arthur Andersen admitting to any wrongdoing.

A RISKY MERGER

While Hills and his team were focused on the financial issues, Miller was quietly working on two tracks. One was focused on recruiting a new CEO, and the second, known to only a few in the company, focused on the possible merger of the company with the much smaller USA Waste Services. Soon after arriving, Miller had gotten a call from Jerry York, a former colleague and friend from his Chrysler days, who was on USA Waste's board and proposed combining the companies. Miller listened and met with USA Waste's top executives. He thought the initial terms discussed were unsatisfactory, and in any case, thought it premature to pursue until the board had completed its financial review. Rumors of a merger had even started to circulate on Wall Street.

USA Waste was an industry consolidator, started four years earlier by BFI veterans John Drury and Rod Proto. They had quickly grown their business through a series of high-profile acquisitions to more than $2.5

billion in annual revenues. The USA Waste management team were still darlings of Wall Street and provided Miller with a ready solution to his CEO search. And USA Waste's singular focus on North American solid waste operations appealed to the activist funds now heavily invested in Waste Management common stock.

The possibility of a Waste Management-USA Waste combination was not without controversy. Buntrock strongly opposed such a merger, and quietly, so did a number of the remaining Waste Management senior officers advising Miller. Buntrock was no longer on the board, but he contacted every director to express his concerns. In Buntrock's view, USA Waste was little more than a roll-up, managed by a deal-oriented management team with no experience with, or appreciation for, the corporate support systems required to manage a business of the size and scope of Waste Management. There were financial management systems, risk management, legal, government affairs, training, and other capabilities needed. They were the resources placed at the corporate office that supported the operations managers in the field and allowed them to serve their customers unencumbered by those additional concerns.

Buntrock communicated his views to the board. USA Waste had grown so fast, essentially doing a rollup just as Waste Management and BFI had in their earliest days. Buntrock knew what it took. To him, the USA Waste team hadn't proven they could run the much larger company, one that required significant infrastructure to manage its billions of dollars of assets. Miller courteously invited Buntrock to voice his doubts to the board. He wanted the board to hear both sides, not just what he had to say.

Experienced insiders found the projected merger-related cost savings touted by the USA Waste management to be speculative at best. The Waste Management senior executives thought one of the proposed merger's primary selling points—$800 million in cost savings—was unrealistic. To them, it was dreamland. COO Joe Holsten decided to conduct his own study and quietly enlisted Larry Galek, a district controller in the West, to assist him.

"You can imagine what an undertaking that would be, looking at every USA Waste operation, how many routes we would take out, how many supervisors, how many route managers, how many salespeople," Holsten recalled. "We did that for every Waste Management operation. I did quite a bit of landfill work myself because I thought that was where the big numbers would be in internalizing the garbage." He kept asking USA Waste's Proto for information on how the company reached its $800-million estimate. Holsten said he never got it. "The closest we could get to their estimates was about $400 million," Holsten remembered.

"I told the board this on the day they voted on the merger agreement," he said. Holsten was accompanied by Bill Hulligan, the most senior officer in Waste Management's day-to-day garbage operations, and acting chief financial officer Don Chappel. Miller pressed Holsten on the issue.

"The only way I could see them even approaching $500 million would have been if they had gone on a very aggressive price increase campaign immediately after the businesses were put together," Holsten said. "Could that be $800 million? I suppose there's a chance but not one I'd bet the farm on." Such a price increase would have relied on the unknown vagaries of the marketplace. Would other companies also increase their rates to keep pace, or would they lower them in an effort to steal Waste Management's customers and gain market share? Such assumptions were risky. Hulligan agreed with Holsten's estimate. The outcome, however, was preordained. Drury and Proto's simplified decentralized management philosophy appealed to Wall Street and would address, at least for the short term, the company's leadership needs. USA Waste Board Member Jerry York, who had made the initial contact proposing the merger, was assigned responsibility for ensuring the synergy savings.

On March 11, 1998, barely more than a week after the financial restatement announcement, the Waste Management and USA Waste Services merger was announced. Each Waste Management share was to be exchanged for 0.725 shares of USA Waste common stock. Waste Management shareholders would own 60 percent of the combined company. It was expected to have a market capitalization of $20 billion. The Waste Management name would be retained, the company headquarters would move to Houston, and Drury and Proto would assume the positions of CEO and COO, respectively. Miller was to be the non-executive chairman for another year and then pass the chairman's title to Drury. The merger closed on July 16, 1998, and the new management team began implementing their very different vision for the combined enterprise, now valued at more than $19 billion.

"I was more interested in getting competent, credible management in place," Miller said. "And I thought the stuff USA Waste did, that lean and mean business model they were running was probably what our shareholders wanted, and I thought that was probably pretty attractive." The Lens Fund's Nell Minow told Dow Jones News Service, "This is what I was hoping for." She told the *Los Angeles Times*, "We are absolutely delighted. This couldn't have solved Waste Management's problems any better. You've now got a company with great assets combined with a company with a great management team."

Soros's Druckenmiller told Dow Jones, "We are ecstatic." He acknowledged that Soros's investment (more than $800 million) in Waste Management had been a "nightmare." He slammed the company's former executives, praised Drury's leadership, and said, "If I didn't own the stock, I'd be buying now." The investors' enthusiasm for the merger would soon prove premature.

THE FALLOUT

The USA Waste team began cutting its way through the former Waste Management infrastructure with a singular focus to loosen the larger company's centralized corporate control and place responsibility and decision-making in the field. They shed operations outside the traditional North American waste services businesses. They slashed corporate-level expenses, including, in particular, the Oak Brook-based accounting and systems center. Only weeks earlier, before the merger, Waste Management had announced a reengineering of its financial and administrative business processes to include marketing, sales, and customer service. The moves were to lower costs, better serve its customers, create a competitive advantage, and position the company to build shareholder value. Now, the half-completed conversion of the Waste Management accounting systems to a new SAP platform, a massive $100-million undertaking to install new enterprise software, was abandoned midstream in favor of moving the operations to the decentralized USA Waste accounting system. The files, systems, and hardware developed and securely stored for years in Oak Brook were tossed. Even millions of dollars of audiovisual slides and training videos found their ends in dumpsters that were taken to the company's own landfills. But they weren't just carving away the fat of the old company, which was expected and planned—and applauded by Wall Street, particularly by those who had promoted the merger—they were cutting into its muscle. The decision would soon haunt them.

The results of these decisions became apparent by the first quarter of 1999 when the newly merged company reported earnings just below Wall Street expectations. By the time the second quarter 1999 earnings were released, the company was forced to admit that its first-quarter financials were incorrect and had to be restated. For the second quarter, the company reported a year-to-date decline in earnings of $250 million. The stock had been about $55 before the news. It now cratered to $34 a share, shaving about $20 billion in the company's market value. Further, the company disclosed that its decentralized accounting system had been unable to keep up with the demands for billing and customer payment accounting, resulting in a significant loss of revenue. The impact of these disclosures was swift: the board quickly dismissed the entire former USA Waste Services senior management team.

THE CLEANUP

Board members Ralph Whitworth, Miller, and Hills were forced to swing back into action. CEO Drury had suffered a seizure on October 31, 1998, at his lake house in Austin, Texas, and had a procedure to remove a tumor in his brain. He was unable to perform, and so the board named the forty-three-year-old Whitworth interim chairman and CEO. Miller agreed to Whitworth's request that he come back as CEO. "To the extent I helped get us into this situation, I feel an obligation to help get us out of it," Miller told *The Wall Street Journal*. Drury did not recover and died on April 13, 2000, after a seventeen-month battle with cancer. He was fifty-six.

The company's announcement of the changes also disclosed plans to divest certain solid waste businesses and sell all or part of the company's international operations, which generated sales of $1.5 billion. Waste Management also launched a comprehensive effort to establish a reliable financial system, particularly after its accounting and IT systems had been questioned. Miller recruited Don Chappel, a senior financial executive working on divestitures in what was now called "Old Waste," to help put the books back together.

Chappel brought in 1000 PricewaterhouseCoopers (PWC) and 300 Andersen accountants to help. PWC had been USA Waste's internal auditor, and Andersen had been the auditor for both Waste Management and USA Waste. The Andersen teams were deployed to corporate and regional offices, and PWC's people were assigned to districts across the continent. The firms didn't have enough people to cover all the territories and began recruiting. They rented out vacant space in the company's Fannin Street tower and brought in rented furniture. Lobby signs directed the PWC and Andersen contractors to their firms' floors. Pizza boxes and ordered-in food containers littered the offices during mealtimes. "It was pretty much a zoo," Chappel remembered. "But it was incredibly helpful. We did get to the bottom of the problem." The cash was eventually posted correctly and accounts reconciled for financial reporting.

John Slocum, the veteran financial executive from Waste Management who had been instrumental in the Jeddah contract and other start-ups, was called in to review the company's receivables and ensure they were collectible. Slocum engaged the staffing firm Robert Half International to provide the finance professionals he needed. "We audited every commercial and rolloff route," Slocum recalled. "We had people go out to the divisions and physically ride every route. We were swarming people." The finance recruits checked to see whether the container was right, whether the billing was accurate, and verified other route information. "From the route audits, we picked up millions of dollars in revenue. Some of the people on the board were surprised we could." The audits helped the company properly account for the work performed.

Asked about the merger's promised $800 million in savings that had been a merger talking point, Miller said, "They probably made some generous assumptions as to what costs they could eliminate, and it turns out they were deadly wrong." He told *The Wall Street Journal's* Jeff Bailey in a February 29, 2000, story that the board had been embarrassed. Hills told him, "I'm not proud of the fact that, in retrospect, we didn't know what the hell was going on." Jerry York, the director who had contacted Miller to propose the merger and was assigned oversight of management for securing the savings, soon departed the board.

THE TURNAROUND CEO

Having started down the road of simplifying the company, the board once again embarked on a CEO search, this time looking for an executive whose experience would pair well with a more operationally focused strategy. By November 1999, they found their man in A. Maurice "Maury" Myers, who was in his third year of a turnaround at Yellow Freight, following stints as an executive at three airlines. His extensive experience in logistics businesses would serve the company well as it refocused on operational efficiencies in its core North American waste operations. "We knew he'd be able to manage a big fleet and get the trucks to run on time," Miller said.

Myers had pored over the Waste Management materials before accepting the offer. "I spent considerable time researching the company," Myers said. "And I discovered that despite all the problems and accounting snafus, New Waste [as they now called the combined company] was still throwing off over $1 billion in annual free cash."

Myers moved quickly, suspending the acquisitions program and recruiting managers he knew and trusted. Among the first were Tom Smith, a computer systems specialist he knew from his stints at America West Airlines and Yellow, and Bill Trubeck, also from Yellow, where he was a senior officer and CFO. He also recruited Larry O'Donnell, who joined as senior vice president, general counsel, and secretary. Myers organized the board and set ground rules. He wanted them to know his views of how they should run things. He earned the board's support and confirmed its decision to exit non-waste-related businesses and the international markets. He also traveled widely to meet employees.

The synergies that Wall Street expected turned out illusory. "The synergies ran up the stock to $60 and they started operating the merged company," Myers remembered. "There weren't any synergies, at least none they were able to get at that point."

By his second year, Myers could report progress. The IT systems failure was repaired, replaced by a capability to support the company's financial processes, including general ledger, accounts payable, asset management, payroll, benefits, and human resource functions. Hundreds of computer servers were consolidated back into a centralized system. Myers introduced a new website and set plans to make it useful for customers. The plan to divest international operations had gone forward and was largely completed, generating $2.6 billion in cash. Recycling was recognized as a key focus area on which the company could capitalize.

By 2001, growth would primarily come from operations. "The previous focus on buying revenue as a primary growth strategy is a thing of the past," Myers told shareholders. Tuck-ins, however, would continue. He reduced the number of suppliers. He cut the number of truck configurations from sixty to fifteen, improving maintenance and reducing costs.

In November 2004, after having served the five years he had originally planned, Myers stepped down and executed his succession plan with the board's election of David Steiner as CEO. Steiner had joined the company in 2000 as deputy general counsel and quickly ascended first to the general counsel position and then to CFO.

Steiner acknowledged the company had come a long way since the chaotic time that greeted Myers on his arrival in 1999. "In 2004, we emerged with finality from the phase known as our turnaround mode," Steiner told shareholders. "The long labor of the past five years has given us a solid foundation on which to build."

THE NATION'S WASTE LEADER

Myers had left Steiner a stronger company. Its array of assets, the collection of which had begun in the early 1970s, sustained it as the industry's North American leader. Across the continent, Waste Management's network in 2004 then included 281 solid waste and five hazardous waste landfills, 381 transfer stations; 17 Wheelabrator Technologies waste-to-energy facilities; and 106 material recovery facilities (13 secondary recovery plants where recyclables were further processed into raw materials). The company still employed 51,000 people in 1,200 locations. It served 22 million customers. Its financials were strong: The company generated more than $1 billion in free cash flow, with revenues of $12.5 billion and earnings of $931 million.

Steiner kept the company focused on efficient operations, controlling costs, and, increasingly, greener sustainability goals that advantaged the company's business and environmentally conscious reputation. He managed it through the difficulties of plunging recycling markets and the 2008-2009 recession and tough economic times. In 2008, he led an effort to acquire Republic Services, Inc., the number-two waste services company launched, ironically, by Waste Management co-founder Wayne Huizenga in 1995. Steiner was ultimately forced to abandon the chase for Republic due to the uncertain 2008 financial markets.

Six years later, in December 2014, Steiner announced the sale of the company's Wheelabrator Technologies subsidiary, the business Buntrock and Rooney acquired in 1990 that made it the leader in waste-to-energy industry. Wheelabrator's seventeen waste-to-energy facilities and four independent power-producing operations were sold to the private equity firm Energy Capital Partners for $1.94 billion. The transaction enabled the company to further refine its focus on its traditional core solid waste business and reduce earnings volatility related to electricity sales, Steiner said. The proceeds went toward repurchasing shares. Steiner led the company for twelve years.

On November 10, 2016, following an eighteen-month succession process, the company announced that Jim Fish had been chosen as its president and CEO. In a press release, Board Chairman Robert Reum said: "Jim has consistently delivered results and has the skills and leadership qualities that make him ideally suited for the role. He has a deep understanding of our strategy, impressive financial and operational acumen, and strong support from employees, customers, and investors." The fifty-four-year-old Jim Fish was now the man in charge of Waste Management.

WASTE MANAGEMENT TODAY

On June 17, 2021, Waste Management celebrated its fiftieth anniversary as a public company. In 1971, the Waste Management leaders believed that reaching $100 million in revenues would be an astounding achievement. They could not yet know what was ahead, the challenges they would face, or the opportunities they would be afforded. At the time of its initial stock offering, the company was launched with about $10 million in revenue, earnings of less than $700,000, and assets of about $3 million. At the end of 2020, the company's revenues exceeded $15 billion, generating profits of about $1.5 billion and a $30 billion network of assets spread across North America that touched every major community.

Just as before, new challenges continue to emerge. Waste Management is endeavoring to improve its recycling systems, employing technology to gain ever more value out of the materials it collects, particularly as climate change is a global priority. And in the years ahead, Waste Management will face—and no doubt embrace—calls for its leadership in waste reduction programs.

In June 2021, the families of the founders gathered in Hinsdale, Illinois, to celebrate the 50th anniversary of its first stock offering and to hear the stories of the company's successes and tales of its builders. In August, company alumni and family numbering in the hundreds gathered in Chicago to celebrate, too. More stories were told and memories shared.

The company had come a long way from the modest origins of the Cicero, Illinois, collection depot. Over time, its founders—the strategist and visionary Dean Buntrock, the aggressive acquisitor Wayne Huizenga, and the operations-focused Larry Beck—would see the company's reach far exceed their original vision. Through the years, the company had taken on huge challenges, succeeded in difficult mobilizations internationally, handled the hardest-to-manage toxic wastes and cleaned up the nation's worst contaminated waste sites, and, of course, performed the day-to-day work of managing much of North America's waste.

So, too, had they faced challenges associated with the growth that Waste Management achieved. In building the company, they had overcome the challenges of creating the leading waste management company in North America. The development of technologies to safely and responsibly manage wastes continues. Science is providing new and revolutionary ways to sustain the environment and new ways to efficiently repurpose recyclables and handle and recover energy from society's discards. Today's waste managers face new challenges. And now, the clock ticks on. Another day is upon it. The company's collection drivers rise before dawn. Its people head to their landfills, gatehouses, material recovery facilities, and administrative offices. The youthful Jim Fish and a new generation of managers are now in charge. Waste Management continues on.

ACKNOWLEDGMENTS

The preparation of this book required the help of numerous people.

Foremost among them was my family. I am deeply grateful for the support and encouragement of my wife, Jeanne, for her assistance throughout this project. She reviewed the narrative literally as the words were drafted and, with common sense, provided the most helpful of suggestions and guidance. Her straightforward questions were invaluable and helped me to focus during the many months of work in front of me. Thank you, Jeanne. Similarly, I am grateful to my children—Nora, Kara, and Colin—for their continuing encouragement as I worked on the book. It was a first experience for all of us. They never stopped inquiring about my progress, and their loving support was more helpful than they can know.

I reached out to scores of Waste Management people, and past and present executives. I conducted more than one hundred interviews, visiting many people on more than one occasion. I am deeply grateful to them for the time they provided and for the insights, information, background, and context they contributed to the writing of this book. Their generosity of time, spirit, and goodwill—indeed, their very voices—enriched what I hope is a lively telling of the Waste Management history.

I am truly indebted to Dean Buntrock, the principal founder of Waste Management, who commissioned the compiling of this book. Dean was particularly interested in capturing the story of the company's fascinating development and achievements and, most importantly, the backgrounds of the key people who built Waste Management into one of North America's leading companies.

I interviewed Dean on a number of occasions, sitting with him for extended periods at his Lake Geneva, Wisconsin, and Hinsdale, Illinois, homes and recording his many recollections. He was incredibly generous with his time and had an amazing recall of the details of the company's forebears, its incorporation in 1968 and public offering in 1971, its earliest expansion days, and its decisions to take on challenges no one would have believed possible. His leadership was evident in all of the company's major developments and challenges along the way. To be sure, Dean was the driver of it all, but he was quick to acknowledge the talented people who surrounded him, most particularly Wayne Huizenga, Phil Rooney, Larry Beck, Don Flynn, and Peter Huizenga. It became clear that no significant decisions were made without his involvement and approval. Although the original assignment was to prepare a book about the company—not about Dean—the company's story simply cannot be told without his voice and without crediting him for the leadership and decision-making he exercised in recruiting the individuals who played the most influential roles in the enterprise.

A key disappointment was not being able to interview two men who also played major roles in the company's founding: Wayne Huizenga and Larry Beck. Huizenga was the co-founder most well-known to the public. He enjoyed a higher public profile in successfully establishing other public companies after his Waste Management days and through his ownership of three professional sports teams in Florida. He died in Fort Lauderdale, Florida, on March 22, 2018. At one point, after being offered condolences on Huizenga's passing, Buntrock said: "He was my best friend, and he would have been your best friend, too." Huizenga was eighty years old. Beck brought operating experience and, through his

Atlas Refusal Disposal company in Chicago's south suburbs, participated in the early purchase of the CID landfill in Chicago. The amiable Beck died in Harbor Springs, Michigan, on June 28, 2018. He was ninety.

I am also very thankful to Phil Rooney, who welcomed me and so many others into the company. His leadership, intelligence, talent, energy, and friendship coursed through every level of the company. His contributions were critical to the company's development, achievements, and success.

It is to my regret that not all with whom I spoke are referenced by name in this book. However, they have my sincerest thanks. Their contributions are present throughout the narrative, if not by name, by implication, description of events, and context. Their insights were invaluable in telling the story. At its largest size in the mid-1990s, Waste Management had 73,000 employees engaged in solid and hazardous waste management activities as well as clean air, clean water, environmental technologies, and international operations across the globe. And the company had other subsidiaries engaged in lesser-known niche businesses. (One example was its little-known marine diving unit that helped plug a hole in a tunnel in the Chicago River that had caused flooding in the city's Loop business and retail center in April 1992.) The company employed many, many officers, directors, and managers along with the people who interacted with customers every day and who, through their heavy lifting, are truly responsible for its success. Thanks to all of you.

I am deeply appreciative of the assistance provided by two of my former colleagues at Waste Management, Herb Getz, retired senior vice president and general counsel of the company, and Bob Keleher, a retired staff director of the company and senior aide to Dean Buntrock and Phil Rooney. They are referenced throughout the book as they carried different responsibilities throughout their careers. These gentlemen could not have been more generous in their support. Herb played a critical role in reviewing the various drafts of chapters as they were being written. He offered comments, background, anecdotes, and insight into all that is in the book. Importantly, his guidance helped me avoid errors and offered the most helpful direction on subjects too numerous to know. Herb, I am in your debt, and you have my sincerest thanks and gratitude. Bob Keleher, too, reviewed the chapter drafts and contributed comments, information, and many stories included in the book. Bob played a key role in many domestic and international projects, including contributing to the company's adventures in Riyadh, Saudi Arabia, and Buenos Aires, Argentina. The insights and information he provided were extremely helpful. He most generously aided me in tracking down key managers long gone into retirement for interviews. Additionally, Bob was quick to provide thoughtful and humorous commentary on a range of topics in the book.

A number of Waste Management executives were also helpful. Jim Trevathan, retired for several years as chief operating officer, regaled me with fascinating stories of his career journey with the company starting with his earliest days in hazardous waste sales in Houston. Jim helped open doors for me to other senior current Waste Management executives, including Chief Executive Officer Jim Fish and Chief Customer Officer Mike Watson. Interviews with Fish and Watson offered greater understanding of the company today, including its strategies, goals, successes, and direction. They are outstanding managers, and I am grateful to them for the time they provided to me. Steve Neff, recently retired as the company's top sports marketing executive, provided invaluable insight into the company's sports branding activities and its former involvement with NASCAR and current involvement with the PGA Tour. A longtime executive, Steve also provided helpful background on the company's sales and marketing functions.

I am also grateful to Waste Management Area Vice President Harry Lamberton, who generously helped me engage with the company's Chicago-area operations staff. His kind encouragement—and that of Mike Watson's from his Houston base—made easier my access to the people who are daily responsible for managing waste in the Chicago area. I am indebted to you. I would like to offer my special gratitude to Senior District Manager Chris Disbrow who welcomed me at the company's Cicero, Illinois, hauling facility and for connecting me with veteran drivers Glenda Schaller and Bob Marcione. A longtime Waste Management manager, Disbrow was generous with his time and provided an up-close view of the details of the company's day-to-day work and customer engagement. I had known Schaller through my consulting work at the company more than a decade ago, and it was a pleasure visiting with her again, catching up on her career, and through her, learning of the changes over time in driver responsibility and expertise. So, too, was Marcione a pleasure to meet. He described in great detail his work, his long career, and the care he took to operate the new technology that is used to operate the big hauling rigs safely and efficiently. All personnel at the Cicero facility have my gratitude.

This is not the first book about Waste Management. Author Timothy Jacobson undertook the task earlier in his 1993 publication of *Waste Management: An American Corporate Success Story.* Jacobson skillfully captured much of the company's early days and, most importantly, how the company's roots as a one-horse hauling company operated by Harm Huizenga at the turn of the century would one day become Waste Management, Inc. Jacobson's work was a helpful resource to me in preparing for interviews with Buntrock and others about the company's early days, its Dutch roots, and initial stock offering and also in generating background to help describe the company's rich early history.

Similarly, I am particularly indebted to Harold Gershowitz, a retired senior vice president of the company whom Buntrock called on in 1966 to work with him and other industry leaders in forming the National Solid Wastes Management Association. Harold was not only a leader in developing the association and giving it a national voice but would be lured to Waste Management by Buntrock and Wayne Huizenga six years later to be its president and handle all matters external. I am indebted to Harold, or Hal as his friends call him, for his generous contribution to the details of this history and, not insignificantly, for playing the key role in launching my own career in Waste Management's communications group in 1986. Thank you, Hal.

I spoke with a number of people involved in Waste Management's initial public offering of stock in 1971. Key among them was Jack McCarthy, retired from the Bell, Boyd & Lloyd law firm in Chicago, who represented the offering's underwriters. McCarthy would go on to be the company's principal outside legal counsel, and I am indebted to him for his assistance. I am also grateful to P.J. Huizenga, the great grandson of Harm Huizenga. P.J.'s father, Peter, was a young lawyer at the time of the offering who played one of the most influential roles in organizing the legal and financial resources needed, and later, as a key officer of the company.

A number of former Waste Management executives provided background on the early days of the company and offered insights into how they came to Waste Management and served the company. Among them were financial officers Jerry Girsch, Bob Paul, Jerry Seegers, Ron Jericho, and Royal Johnson. Many others gave insight into the company's beginnings, including Tom Frank, Earl Eberlin, Dick Molenhouse, Jim O'Connor, Gene Price, and Brian Oetzel. John Melk, describing early acquisition efforts, was remarkably helpful.

I am also grateful to Don Price for his assistance and long friendship. Don, who was Buntrock's second hire and a career officer and enjoys a detailed memory of those early days, offered tremendous insights into the company's history. He was involved in numerous company ventures and provided valuable details on the company's development ranging from early acquisitions and foreign mobilizations to the start of Chemical Waste Management and hauling operations over two decades. Don had retained a variety of materials from his career, including decades of work calendars, directories of employees and trade associations, photographs, and other materials that he generously shared with me for the preparation of this book. Don, you have my sincerest thanks.

Similarly, Bill Debes offered enormous help in detailing the company's approach to its early acquisitions, its financial and accounting systems, and the people required to quickly assess acquisition candidates and integrate them into the enterprise. He explained pooling-of-interests accounting, which aided the company's growth and added to its value, and how he recruited a number of accounting people who enjoyed long careers with the company, including Mike Rogan, Fred Weinert, and Tom Collins.

Others contributed as well. Among them are Bill Hulligan, Jerry Girsch, Earl Eberlin, Jim O'Connor, Jerry Kruszka, Bob Damico, and Dean VanderBaan. Jodie Bernstein, too, was generous in her assistance, recalling her introduction and recruitment to the company at a difficult time in 1983. Many who were involved in government affairs in Washington, DC, and at the state and local level also contributed, including Jack Schramm, Frank Moore, Jim Banks, Sue Briggum, Chuck McDermott, Ed Skernolis, Kevin Igli, Leah Haygood, Ron Hogan, and Kent Stoddard.

Lou Waters, one of the founders of Waste Management's rival, Browning-Ferris Industries, also agreed to an interview, and I am appreciative of his generosity in visiting with me.

I am likewise appreciative of the enormous help and cooperation I received from executives who led the international mobilizations in Riyadh and Jeddah, Saudi Arabia; Buenos Aires, Argentina; and those who undertook acquisitions across Europe, Australia, New Zealand, and Asia. John Melk, Fred Weinert, Ed Falkman, Mike Rogan, John Blew, Tom Frank, Jim McGrath, Don Price, Bob Keleher, Bud Ingalls, and John Slocum all contributed to my reporting the details, complexities, and tales of the Saudi Arabian adventures and how, though challenged, they achieved their amazing successes. A number of them were also involved in the Buenos Aires project and contributed background and insights, including Weinert, Falkman, Keleher, and Debes, and also Jerry Girsch, Alex Gonzalez, and Bob Van Tholen. My thanks also go to Frank Schroeder and Joe Holsten, who previously headed Waste Management International, for their help in gathering material on the company's expansion across Europe.

I had the help of Don Price, Ray Bock, Jim Koenig, Bob Keleher, and others in gathering material on the development of the company's Chemical Waste Management and remediation businesses. Gaining an understanding of operations in Pennsylvania required the assistance of Bill Hulligan and Joe Holsten. My sincerest thanks are extended to Dennis Grimm, Jane Witheridge, and Peter Yaffe for their valuable insights into the company's expansion in Pennsylvania.

Witheridge, Bill Moore, Stu Clark, Bob Russell, and Kent Stoddard offered background on Waste Management's involvement in recycling. They were in the trenches, and their recollections were invaluable. Herb Getz and Jim Koenig provided many of the details on the company's involvement and later majority interest in Wheelabrator Technologies.

In addition to those who contributed to the writing of this book, I am particularly grateful to the talented people in the Oak Brook communications and public affairs group with whom it was my privilege to work. They included Don Reddicliffe, who was instrumental in bringing me into the company, and Joe Pokorny. Other talented communicators through the years included Chris Collins, Sharon McIntosh, Laura Field, Dianne Brooks, Monica Linsin, Bob Reincke, Michael McCarthy, Janelle Jones, Mary Orr, Rebecca Landy, Vivian Pearson, and Kevin Belgrade. The group was supported by a talented audiovisual department headed by Jim Benefield that included Arlene May, photographer Charles Hopkins, and producer Roy Cohen, the latter leaders in creating company presentations and launching an internal satellite television system that stretched across the continent to reach hundreds of operating locations. I worked closely with other talented professionals, including those in the London office, Peter Coombes, Tom Henderson, and Jackie Jacobie, as well as Kevin Stickney and Patti Powers at Wheelabrator's headquarters in New Hampshire. Two others played significant roles in the group: Ron Shufflebotham, the longtime director of public relations affectionately known as "Shuff," was responsible for, among many, many things, branding and staging of trade shows and the annual stockholders' meeting. His plans were detailed and voluminous. He thought of everything, he forgot nothing. Don O'Toole was director of corporate advertising and, working with Bill Whitney's Ogilvy & Mather Chicago advertising shop, created award-winning commercials. As a consultant to the company, I was privileged to work with Steve Batchelor, Barry Caldwell, Lynn Brown, Sarah Simpson, and the communications group in Houston.

The Oak Brook communications group enjoyed a talented team of aides. Mine was Darlene Chesky. Darlene was my administrative partner for more than a decade and looked out for me in ways I am confident I don't even know. She always understood the priorities and where a day's energy should be invested. Darlene was always in the middle of whatever communications initiative or announcement was needed. Indeed, she worked closely with the financial team, and it was through her labor and commitment that hundreds of press, financial, and internal announcements flawlessly flowed through the years. Darlene was the ultimate professional.

Members of the company's legal team also assisted. They included Tim Casgar, the company's first general counsel; Jodie Bernstein, recruited by then-general counsel Steve Bergerson in 1983 in the midst of a media crisis; and, of course, Herb Getz. Waste Management had a stellar legal team composed of more than 150 lawyers, many of whom I worked closely with during my time as the company's vice president of communications during the 1990s. The issues were numerous, varied, and global, requiring close coordination between the communications department and the expert legal staff. My recollections of these people and the many issues on which we interfaced contributed to my institutional memory running throughout this history. They included a number of attorneys at the corporate office including Dick Houpt, Mike O'Brien, Jim Hynes, Brian Clarke, Dave Coleman, Tom Witt, Roger Berres, Greg Constantino, Ian Bird, Tom Genovese, John Noel, John Ray, Jeff Evert, Brett Heinrich, Brian Blankfield, and Diane Dygert. Also Vaughn Hooks, Peter Kelly, Andi Kenney, Alyse Lasser, Jan Reed, Ray Martinez, Greg Seidor, Greg Sangalis, Steve Stanczak, Ernie Summers, Tom DeMay, Dale Tauke, Linda Witte, and Lisa Zebovitz. There were frequent matters coordinated with attorneys in the company's regional offices, including Dave Domzal, Carl Frank, Bill Jeffry, Tom Jennings, Marian King, Bob Leininger, Debbie Romanello, John Skoutelas, Mike Slattery, John VanGessell, Dennis Wilt,

Jack Cassari, and Howard Yamaguchi.

I am equally grateful to the many corporate financial managers who were colleagues. The list is long and regretfully incomplete, but key among them were Jim Koenig, Tom Hau, John Sanford, Don Chappel, Mark Spears, Ron Jericho, Bruce Tobecksen, Bill Keightley, Cherie Rice, Sue Nustra, John Repke, and John Marek.

I offer my sincerest thanks to those many, many colleagues with whom I worked but who are not referenced by name in these pages. It has been many years since we were together, and you have my gratitude for your support and friendship over the years.

I am also grateful to former Waste Management CEOs Steve Miller and Maury Myers for generously providing me with opportunities for interviews about their roles in the company and their involvement during challenging periods. They were generous in their time and candid in their comments.

Finally, I offer my gratitude to Rosemary Lane, who energetically took on the task of editing this book. Her editorial suggestions, questions, and clarifications have contributed enormously to the final product. I am indebted to Rosemary for applying her editorial craft to improve this narrative's flow, readability, and accuracy. Rosemary is a true professional and has my sincerest thanks.

—WILLIAM J. PLUNKETT, MAY 2022

INDEX

Virginia, 104, 207
VOC companies, 290
Vulcanus I and *II*, 198, 214

W
Waco, Texas, 279
Wald, Harkrader & Ross, 181
Wald, Pat McGowan, 181
Walker, Dave, 172
Wall Street, 89, 92–93, 125, 180, 181, 183–
    184, 187, 299, 300, 301, 303, 306
Wall Street Journal, the, 188, 241, 242, 277,
    285, 304, 305
Wallace, George, Gov., 191, 231
Wallgren, Don, 99–100, 202, 204, 265
Walmart, 281
Warner Company, the, 251
Washington, DC, 20, 38, 68, 122, 200, 201
Washington Post, 144, 190, 197, 234
Waste Connections, 10
waste industry consolidation, 59
waste industry's convention, 153
Waste Management,
    401(k), 15
    1970 revenue, 50
    1991 revenues, 273
    2020 revenues, 8
    accounting, 69, 116
    accusations against, 179–184
    acquisition playbook, 75, 216
    acquisitions, 10, 58, 59, 61, 63, 65–72,
        75–86, 103–123, 262, 272
    acquisitions, non-waste, 274–275
    adviser on Superfund, 201
    allegations about CID site, 242
    analytics, 12
    assets, 8, 10
    Avenue A transfer station, 99

    base in London, 214
    becoming industry recycling leader,
        259–271
    beginnings, 10
    BFI and, 48, 69, 81, 84, 85, 90, 109
    board, 8
    brand platform, 12
    branding change, 14
    British campaign, 219
    Buenos Aires project, 153–166
    burning waste at sea, 233–234
    business development, 58
    business ethics program, 183
    business goals, 15
    business in Argentina, 166
    businesses in Indiana, Minnesota, and
        Ohio, 71
    business model, 10
    CEO, 8, 10
    CFO, 232
    changes in CEO, 293–297, 308
    chemical waste facilities, 182
    chemical waste operations/unit, 150,
        199
    chemical wastes management, 138
    CID (Calumet Industrial District)
        facility, 46, 93, 96, 149, 228, 230,
        242
    collection company in France, 218
    collection trucks, 72
    commercial accounts, 72
    commercial customers, 103
    community relations campaign, 164,
        232
    computer programming and systems, 69
    consolidated financial statements, 58
    controller(s), 116, 120–121, 220
    corporate spokesman, 92
    cost-cutting program, 284